Columbia University

Contributions to Education

Teachers College Series

No. 400

AMS PRESS
NEW YORK

A STUDY OF BIOLOGY NOTEBOOK WORK IN NEW YORK STATE

By

DON O. BAIRD, Ph.D.

Teachers College, Columbia University
Contributions to Education, No. 400

Bureau of Publications
Teachers College, Columbia University
NEW YORK CITY

1929

Library of Congress Cataloging in Publication Data

Baird, Don Otto, 1888-
A study of biology notebook work in New York State.

Reprint of the 1929 ed., issued in series: Teachers College, Columbia University. Contributions to education, no. 400.
Originally presented as the author's thesis, Columbia.
Bibliography: p.
1. Biology--Study and teaching (Secondary)--New York (State) 2. Note-taking. I. Title. II. Series: Columbia University. Teachers College. Contributions to education, no. 400.
QH315.B3 1972 574'.071 71-176532

ISBN 0-404-55400-8

Reprinted by Special Arrangement with Teachers College Press, New York, New York

From the edition of 1929, New York
First AMS edition published in 1972
Manufactured in the United States

AMS PRESS, INC.
NEW YORK, N. Y. 10003

ACKNOWLEDGMENTS

The author wishes to express his appreciation to his dissertation committee. He is especially indebted to Professor S. R. Powers, chairman, who suggested the problem and made it possible for the study to be undertaken. Grateful acknowledgment is due to Professor Otis W. Caldwell and Professor William H. Kilpatrick for their helpful suggestions and criticisms, and for the inspiration received in their classrooms.

Credit is also due to C. E. Baer, formerly State Supervisor of Science of New York State, for his aid in securing the notebooks from the cooperating schools. The help and criticism of Professor L. A. Waits, of Cincinnati, Ohio, in the preparation of this manuscript are deeply appreciated.

Thanks are especially due to the teachers, principals, and pupils of the selected schools for their splendid coöperation and ready response to the questionnaire study.

D. O. B.

CONTENTS

TABLES

CHARTS AND MAP

CHAPTER I

INTRODUCTION

The Problem Defined

This study is an analysis and evaluation of biology laboratory notebook work. Each notebook used in this study has been accepted by the teacher as a record of a year's work in laboratory experiments, exercises, and studies for credit in high school biology.

S. R. Powers,[1] in a recent article in the *Teachers College Record,* says:

> The problems pertaining to the laboratory notebook are similar to those which relate to the laboratory itself. It must be true, if the notebook has any place in science teaching, that its place is justified by the fact that it contributes learning experiences which are helpful for accomplishment of definable objectives. There is need for research which will determine what the possible learning experiences are, and for research that will show how these experiences may be made real for the learner.

The purpose of this study is to determine, through an analysis of representative biology notebooks, the contribution of biology notebooks of the secondary schools of New York State toward a realization of the biological and general educational objectives in education, and to evaluate scientifically these findings. This study also attempts to reveal some of the factors that underlie the present trend in notebook practice; and for a clearer understanding of this a study of the closely related factors, such as texts, laboratory manuals, courses of study, examination questions, and college entrance requirements is included.

Historical Sketch

Pestalozzi (1746-1827) is credited with being one of the earliest educators to introduce (1776) living objects as materials of interest in the instruction of children. In 1842 Harvard College

[1] S. R. Powers, "Research in Science Teaching," *Teachers College Record,* January, 1929, pp. 334-342.

appointed as a professor of natural history, Asa Gray (1810-1888), who was recognized as the foremost botanist of his time in America. Louis Agassiz (1807-1873), in 1848, also was made a professor of natural history at Harvard College, and he developed in this institution the first laboratory and field course in the study of animals. Agassiz's motto "Study nature, not books" had much to do with the popularizing of biological laboratory work.

In its modern sense, biology did not appear in the curriculum until about 1860, when the unity and nature of plant and animal protoplasm were first clearly demonstrated by Max Schultze (1825-1874). For this evidence of the likeness and fundamental relationship of living things Schultze is called the father of modern biology. Credit must also be given to Lamark (1744-1829) and to Treviranus (1776-1837) for establishing the unity of life and for coining the word "biology" about 1803. Biology as a unified science of living things was brought to this country in 1876 by Huxley.

Downing in his recent book, *Teaching Science in the Schools,* notes that: It was the academies of New England and the Middle States that first began to offer credit courses in botany and zoology. Botany was offered in the academies as early as 1795 and zoology in 1818. The University of Michigan accepted both botany and zoology as entrance requirements in 1873. Botany as a high school subject appeared in the Boston High School for girls in 1826; human physiology at Ipswich in 1839; and zoology at Waltham in 1849. Biology as a unified subject is a very recent science in the public school curriculum, being preceded by botany, zoology, and human physiology.

Biology first appeared in the high schools of New York State about 1905, followed by the first Regents examination in this subject in June, 1906. The College Entrance Board offered the first examination in biology in 1914.

The use of notebooks for the keeping of records of science discoveries goes back to antiquity—to the time of the Babylonians and the Egyptians. Notebooks in biology followed the use of the herbarium for botany and of the drawing for zoology. These science subjects in turn inherited notebook practices from the older formalities in notebook work in chemistry and physics. Along with all that was good in notebook practices in these subjects was inherited also all that was evil. The present tendency favors laboratory demonstrations with note-taking followed by the writing up of such demonstrations in a permanent record in the notebook.

Procedure

A careful search of the literature in the field has failed to reveal any specific study of the analysis and evaluation of biology laboratory notebook work. In view of the lack of definite information on this phase of educational work, it has seemed worth while to the author to undertake this study.

Fifty-two notebooks were collected from schools geographically distributed throughout the state of New York. These schools are representative of small, medium, and large secondary schools, both public and private. The notebooks were critically examined as to the nature and number of experiments; kind and amount of drawing; extent and comprehensiveness of pupil vocabulary; quality of handwriting; amount and kind of health material presented; the use of leisure time; etc.

The textbooks used in these schools were analyzed as contributing factors to laboratory and notebook work; likewise the laboratory manuals and teachers' guides. A special study was made of the Regents and College Entrance Examination Board's examination questions over a period of years to determine what, if any, influence such questions have on notebook work. The scientific literature, as found in science and research magazines, was carefully searched for ideas in the field of biology pertaining to biology notebook work.

Finally, questionnaires were sent to the biology teachers in the cooperating schools for their collective judgment and evaluation of the findings; to the principals for data on both pupil and teacher, as well as on school facilities for the teaching of biology adequately; and to the pupils who contributed their notebooks, for their ideas of, and reactions to, the value and personal interest of biology notebook work to them.

The geographic distribution of the cooperating schools of the State of New York is shown in Map No. 1 and a table of the names and populations of the cities and villages in which the schools are located (Tables 1A and 1B).

It is of course not claimed that the items chosen for this study

TABLE 1A

POPULATION OF TOWNS AND CITIES IN WHICH COOPERATING SCHOOLS ARE LOCATED*

Location	Population	Location	Population
Albany	117,820	Mamaroneck	10,153
Ballston Spa	4,476	Medina	6,192
Batavia	15,628	Middletown	20,412
Binghamton	71,915	Mt. Vernon	50,382
Brockport	3,626	New Paltz	1,269
Buffalo	538,016	Niagara Falls	57,033
Corning	15,722	Norwich	8,345
Courtland	13,879	Oneonta	12,057
Depew	6,122	Oswego	22,369
Dunkirk	19,912	Oyster Bay†	3,700
Far Rockaway		Perry	4,636
Geneva	15,908	Plattsburgh	11,552
Glens Falls	17,851	Pleasantville	3,674
Gouverneur	4,111	Poughkeepsie	35,670
Haverstraw	5,925	Rochester	316,786
Hornell	15,784	Salamanca	9,878
Ilion	10,426	Saranac Lake	6,579
Ithaca	18,948	Saratoga Springs	13,884
Jamestown	43,414	Schenectady	92,786
Johnstown	10,712	Scotia	5,562
Kingston	28,099	Syracuse	182,003
Lima	981	Tarrytown	6,199
Lockport	21,676	Utica	101,604
Lyons	4,275	Watertown	32,836

* Population figures taken from the New York (state) Census, 1925.
† Population in 1920.

TABLE 1B

CLASSIFICATION OF PLACES IN TABLE 1A AS TO SIZE

Total Number of Schools	Population							
	Less than 1,000	1,000 to 5,000	5,001 to 10,000	10,001 to 15,000	15,001 to 25,000	25,001 to 50,000	50,001 to 100,000	Over 100,000
Total (48)	1	8	9	7	10	4	4	5

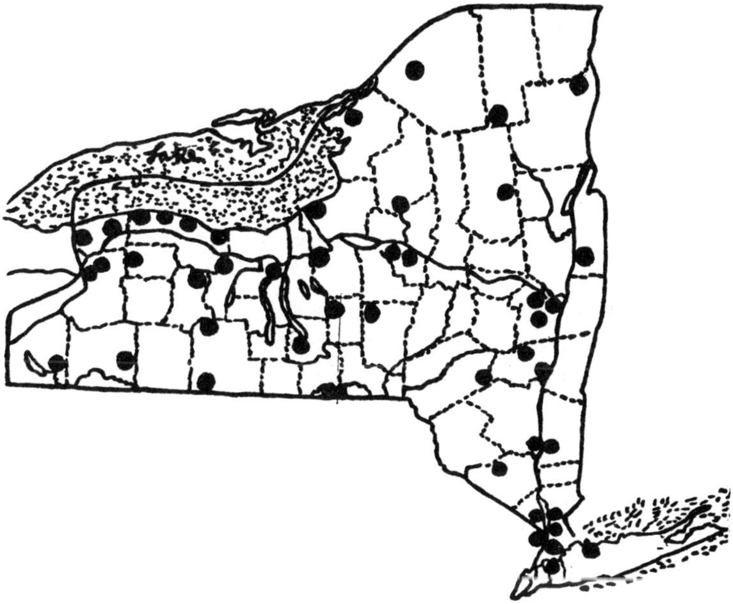

MAP NO. 1

GEOGRAPHIC DISTRIBUTION OF THE COOPERATING SCHOOLS OF THE STATE OF NEW YORK

exhaust the subject. In particular it is recognized that many other aspects of notebook work are probably more important, could they be objectively treated, than those selected. The items chosen, however, are deemed important and have the advantage of lending themselves readily to objective and quantitative study.

CHAPTER II

THE WRITTEN UNITS OF WORK

It was necessary to work out some definite plan for the study and evaluation of the notebooks. After reading most of the notebooks to gain a general idea of their contents, the following procedure was adopted:

Five representative units of written work (Oxygen, the Frog, Osmosis, Pulse Rate, and Photosynthesis) were examined for the following points—number of pupils reporting each of these five units of written work; total number of words used; range of words used; and total number of separate words. Then a study was made of all the units of written work in all the notebooks, as follows: number of words per page, estimated total number of words, percentage of written matter and of drawings; number of pages in each notebook, size of the pages; and any unusual characteristics. Another study was based on the number of units of written work in each notebook, number of weeks in the course, names of the textbooks and laboratory manuals used with the names of the authors, and sex of the teachers and pupils. The drawings were analyzed as to type—whether analytical or non-analytical, pen, ink, or crayon; the total number of drawings in each notebook were counted; and unusual types were noted. Finally, a study was made of the subject matter classified as: chemical and food, plants; lower animal, human physiology, and mixed or miscellaneous material. A special study was made of the health material content of the notebooks.

In estimating the number of words in a notebook, the words on several pages were counted, an average was taken, and this number was multiplied by the number of pages in the notebook. Since all pages in the notebooks were not full pages and most of them combined drawings with written matter, a rule was used to determine the proportion of drawings and written matter in terms of percentage of space devoted to each in every notebook.

The percentage of written matter and of drawings in the notebooks was computed as follows: The number of pages in the notebook was divided into the number of pages of drawings. This gave the percentage of drawings, and the difference between this per cent and one hundred per cent gave the percentage of written matter.

In the word count everything written by the pupil was counted, whether the word or words were applicable or not. Each number was counted as one word; for example, 7 was counted as a word and also 163, 13, etc. Such words as Purpose, Materials, Observation, etc., designating divisions of an experiment, were also counted. Words written in by the teacher for corrections were not counted. Statements like "Experiment," "28," dates, and the pupil's name in connection with individual units of written work were not counted.

In order to verify the count of words and to insure reasonable accuracy, the words in one of the larger notebooks, in one of the smaller, and in two of average size were actually counted. This verification showed that the estimated number of words in the notebooks was approximately correct. In the study of the drawings every separate part, if separately labeled, was considered as a drawing, even though it was evident that two or more parts really constituted a complete study of a single unit. All written or hand printed words pertaining to the drawings were counted in estimating the total number of pupil words in the notebook. For example, if the drawing had a title, the words of the title were included in the total count of words, as were also words used as labels.

In the case of the printed laboratory notebooks all words used by the author in connection with the experiments were counted, even though some words were part of the references rather than a necessary part of any particular experiment, exercise, or study. The number of words used by the author in the printed type of notebooks was estimated by counting the words on several pages, taking the average, and multiplying by the total number of pages of printed matter in the notebook. This gave a rough estimate of the total number of printed words.

This study of notebooks classifies all experiments, exercises, studies, etc., as units of written work. It was necessary to do this because of the difficulty of determining just what material in the notebooks could be definitely placed as an experiment, a problem, an exercise, a study, etc. The same difficulty was encountered in the attempt to classify the drawings. It was quite impossible to designate and group the drawings as sketches, diagrams, drawings, etc. Thus the simplest plan was to consider each separately labeled graphic representation a drawing.

In the vocabulary studies one of the smaller and one of the larger notebooks were selected and all the words were listed alphabetically, counted, and checked against Thorndike's *Teacher's Word Book* and Powers' *List of Scientific Terms for High School Students.* The same procedure was then followed with two of the medium-sized notebooks. The vocabulary obtained from these four notebooks was then used as a check list against the remaining forty-eight notebooks for additional words not found in the first four, and the whole vocabulary of the fifty-two notebooks was arranged as a vocabulary of the total separate words in the notebooks. The words in this vocabulary not found in the Thorndike list or the Powers list are arranged in alphabetical order in Table 11.

The handwriting in each notebook, except two that were hand printed and one that was typewritten, was measured with Ayres' Handwriting Scale, Gettysburg edition.

The notebooks were also examined for the kind and amount of health material in them, and for material of value in prevocational training and training for the worthy use of leisure.

After the study of the notebooks had been completed, the units of written work were classified as to subject matter to form the basis for questionnaires to teachers, principals, and pupils. The questionnaire to teachers consisted of two parts, one of which called for a specific evaluation of the kind and type of material found in the notebooks. The questionnaire to the pupils attempted to obtain their reaction to the kind of notebook work they had been required to do for credit in their biology course.

The fifty-two notebooks for this study were collected from forty-eight cities and villages in New York State. A teacher of biology in each of the cooperating schools had been requested by the New York State Department of Education to select for this study a good but not the best notebook from all the notebooks of their biology pupils. Ten boys and forty-two girls submitted notebooks. Forty-two of the teachers of these boys and girls were women. The general statistical data evolved from a study of the notebooks are given in the tables which follow and are discussed in connection with the tables. The geographic distribution of the schools operating in this study is shown on Map 1. The names of the cities and villages with population data are given in Table 1A. A list of the textbooks and laboratory manuals used in the cooperating schools is given in Chapter V.

A preliminary study showed that an average notebook contains a total of 9,401 words, that the average number of units of written work is 46, and that the percentage of written matter and drawings is 71 and 29 per cent, respectively. It was also found that the average notebook contains units of written work distributed as follows: Plant, 19; lower animal, 7; human physiology, 5; chemical and food, 10; and mixed or miscellaneous, 2. The remaining units of written work consists of bird charts, essays on birds and bird laws, and biographical sketches of men of science. No single notebook, however, agreed with the mythical average.

Generally considered, the units of written work may be classified as experiments to test for something such as a test for grape sugar, and experiments to observe the actions or functional nature of something, such as the study of a living frog or fish. On this basis the average notebook contains 12 units of written work of the first type and 16 of the second type. The remaining units of written work are not classifiable in this manner. Most of the units of written work are relatively short, but in every notebook there are one or more studies of an extended type, such as the study of the frog, the fish, or the grasshopper. In all the notebooks the emphasis is on plants rather than animals, and but little effort is made to show the functional relationship of plants and animals to man.

The total number of words used by pupils in writing the fifty-two notebooks is estimated as 488,826. The total number of separate words in all the notebooks is 2,358. The number of separate words more or less distinctly scientific terms is 645. One of the smaller notebooks contains 422 separate words, while one of the larger ones contains 1,781.

A Cross-sectional View of the Notebooks

Five typical units of written work were selected for intensive study, namely: Oxygen, Osmosis, the Frog, Pulse Rate, and

Photosynthesis. These units of written work were studied for the following points: number of pupils reporting them; total number of words used by each pupil reporting them; range of the words used; and number of words used as label words for drawings.

These five typical units of written work were also selected to represent a cross-sectional view of the notebooks because they were reported by the greatest number of pupils submitting notebooks, and because they may be classified as to subject matter as follows:

Chemical.............................Oxygen
Food.................................Osmosis
Plant................................Photosynthesis
Lower animal.........................The Frog
Human physiology.....................Pulse Rate

The data obtained from the notebooks on these five units of written work are tabulated in Table 2.

TABLE 2

UNITS OF WRITTEN WORK SELECTED TO GIVE A CROSS-SECTIONAL VIEW OF NOTEBOOK WORK

Written Unit of Work	Number of Pupils	Total Words	Average Number of Words	Range of Words	Number of Words in Labels	Average Number of Words in Labels
Oxygen	48	8,236	176	30– 441	332	14
Osmosis	48	8,144	170	50– 370	654	17
Photosynthesis ..	39	4,930	126	36– 212	95	2
Pulse Rate	28	2,360	84	3– 199	0	0
The Frog	43	19,022	488	33–1,262	1,699	47
Totals..........	206	42,692	1,044	152–2,484	2,780	80
Average	42	8,538	209	30– 497	556	16

In the case of the study of the frog, everything the pupil had in his notebook concerning this animal was considered, for the purpose of this investigation, as one unit of written work, even though two or more divisions or separate units were outlined. This explains, in part, why the data show pupils used so many more words in writing the record of the work on the frog than in recording the work on oxygen, osmosis, and the rest. The average number of words used by pupils in recording their observation and study of the frog is 488. The range of total words used is wide—33 to 1,262. The wide range in the use of words is due, largely, to the fact that many of the printed forms of laboratory manuals require merely the filling in of data.

In the study of the photosynthesis unit of written work the chlorophyll type unit was selected, except in cases where only light or a combination light and chlorophyll was reported in the notebook. The oxygen unit of written work is the only unit of this group found in all the notebooks, but in four of the notebooks it varied so much from the usual type that it could not be included in the comparative table. The average number of words used by pupils in recording the oxygen unit is 176. The range of words used is from 30 to 441.

The pulse rate is not a good example to represent the human phase of biology notebook work, but in a consideration of the material in the notebooks this unit appeared most often.

The units of written work in the notebooks were also classified as to general types—a test for something, such as the chemical type of exercise, and an experiencing or observing of the functional nature of something, such as a study of the living frog or grasshopper. The total number of units of written work of these two general types is 1,600 of which 645 units are of the type to test something and 955 units to observe the functional nature of something. Specimens of these two types of exercises are shown on pages 31, 32, 33, 35, 36.

There are relatively few units of written work of an extended type, such as the study of the frog, and no work covering a long period of time such as the breeding of guinea pigs to develop a long-haired breed or a short-haired breed; the use of an aquarium to show the effect of reducing or increasing the relative proportion of plant and animal life, etc. The nearest approach to the extended type of written units of work is the work on the frog, the grasshopper, and the fish. There are, in many of the notebooks, considerable data on flowers, the influence of alcohol and tobacco on the human body, the study of birds, and the conservation of forests, but most of this work consists of essays and copied material.

The total number of units of written work in all the notebooks is given in Table 3.

TABLE 3

Number of Units of Written Work in Each Notebook, Including Bird Charts, Biographies, Etc.

Number of Units of Written Work			
71	22	51	36
49	47	50	25
31	32	52	38
36	41	42	83
29	42	48	25
57	63	73	53
40	96	42	40
29	34	67	52
60	34	43	56
42	40	66	41
66	47	37	
55	46	35	Totals 2,397
24	21	53	Average 46.09
36	52	47	

The range in number of units of written work in the notebooks is from 21 to 96, and very few notebooks have the same or even approximately the same number of units. It is evident that the amount of work represented by the material in the notebook for credit for a year's course in biology must be far from uniform. In one case a notebook containing 21 units of written work is accepted as meeting the requirement for a record of the laboratory work of the secondary school course in biology, and in another case a notebook containing 96 units of written work is presented and accepted as a record of the work done in the laboratory. Between these two extremes is a great variety of number of units of written work recorded in the notebooks and accepted by teachers as fulfilling the biology notebook requirement in their respective schools. This situation would seem to emphasize the need of some standardization of notebook work in secondary schools offering courses in biology.

General Statistics in the Notebooks

The distribution of the units of written work in the fifty-two notebooks, classified as to subject matter, but omitting units of written work on bird charts, biographical sketches, and other material not included in exercises, is presented in Table 4.

TABLE 4

DISTRIBUTION OF THE UNITS OF WRITTEN WORK IN THE NOTEBOOKS

Total Plants	Total Lower Animal	Total Chemical and Food	Total Human Physiology	Total Mixed or Miscellaneous	Grand Totals in Each Notebook
31	24	12	7	2	76
22	4	13	1	1	41
22	6	15	3	1	47
18	5	14	2	1	40
12	9	10	3	1	35
22	3	10	3	1	39
28	5	14	2	1	50
23	7	14	1	1	46
13	9	11	2	1	36
18	12	10	5	7	52
31	11	9	9	4	64
29	9	15	1	2	56
14	8	11	7	2	42
16	1	6	1	0	24
16	5	12	2	1	36
15	3	11	2	4	35
16	1	16	2	1	36
5	1	12	2	5	25
18	0	10	0	5	33
26	5	10	9	3	53
11	1	8	0	2	22
16	15	9	4	3	47
15	4	9	3	1	32
15	11	12	8	4	50
16	0	6	2	2	26
17	5	4	6	2	34
32	8	11	10	3	64
23	8	13	3	6	53
35	28	12	7	2	84
7	1	7	5	2	22
20	11	12	4	1	48
17	10	4	8	2	41
10	6	9	8	3	36
15	1	14	3	3	36
23	13	14	4	4	58
16	6	15	3	2	42
30	10	0	9	3	52
22	0	6	0	2	30
19	*6*	*10*	*5*	*4*	*44*
15	6	10	4	2	37
13	9	10	5	2	39
26	12	4	9	4	55
10	2	5	15	1	33
16	12	4	6	0	38

TABLE 4—*Continued*

Total Plants	Total Lower Animal	Total Chemical and Food	Total Human Physiology	Total Mixed or Miscellaneous	Grand Totals in Each Notebook
18	0	12	3	2	35
14	4	14	9	2	43
15	7	5	3	3	33
23	5	10	3	2	43
22	9	3	12	2	48
16	1	2	4	2	25
20	18	13	6	7	64
29	11	13	20	5	78
991*	368*	515*	255*	129*	2,258*
19†	7†	10†	5†	2†	43†

* Total. † Average.

The data in the foregoing table may be interpreted to reveal a number of interesting things, but most significant is the lack of uniformity in the amount of work presented in a biology notebook. It is also important to note the emphasis placed on plants in laboratory work. Nearly four times as much work is devoted to the study of plants as to the study of animals and the human phase of biology. The study of chemistry and foods in the biology laboratory constitutes nearly a fourth of the total number of units of written work in the notebooks. No notebook has the units of written work distributed to agree with the averages obtained from Table 4, although one notebook does agree fairly well (indicated in italic.) The units of written work listed as mixed or miscellaneous are units on plant and animal cells written up as a single laboratory exercise and similar work that could not be listed under the first four headings of the table.

The notebooks are of two types—the written type and the printed type. Thirty-nine of the notebooks are of the written type and thirteen of the printed type. There are enough differences between a type of notebook that is printed with space left blank for the filling in of data and a notebook written entirely by the pupil to justify the tabulation of data from each type in two separate tables. Table 5A gives the data on the percentages of drawings and written matter in the notebooks

TABLE 5A

THE WRITTEN TYPE NOTEBOOK. NUMERICAL DATA CONCERNING PAGE SPACE OF DRAWINGS AND WPITTEN MATTER, AND VOCABULARIES

Number of Words per Page	Total Number of Words	Per Cent of Written Matter	Per Cent of Drawings	Number of Pages in the Index	Number of Pages in the Notebook
153	26,148	78	22	4	216
130	3,840	56	54	4	116
127	8,604	60	40	8	113
121	14,520	80	20	3	150
151	33,401	70	30	6	316
150	19,267	67	33	8	386*
122	5,418	85	15	2	51
81	15,613	61	39	4½	316
138	13,938	60	40	4	324*
85	13,770	85	15	3½	192*
76	4,560	71	29	2½	120*
113	16,498	73	27	3	200
140	10,500	78	22	2	92
144	23,184	72	28	9	207*
112	5,712	66	34	2½	152*
105	19,425	82	18	3	210*
67	3,283	80	20	2	122*
132	5,016	64	36	3½	120*
133	17,290	82	18	3	210†
138	9,108	68	32	3½	194*
97	14,550	87	13	6½	230†
64	3,712	68	32	0	114†
120	3,960	55	45	2	120*
102	3,570	64	36	2½	110*
116	5,220	68	32	3	132*
68	4,862	80	20	2	89*
111	5,772	72	28	3	144*
172	14,620	80	20	6½	212*
92	6,762	70	30	3	210*
137	8,220	73	27	3½	164*
97	3,492	60	40	4	118*
71	3,266	62	38	3	150‡
103	7,971	77	23	2	194*
116	2,204	83	17	4½	230*
95	18,156	89	11	5	228*
221	6,182	53	47	2	124§
142	11,216	68	32	0	117
128	4,736	69	31	2½	136*
88	2,812	77	23	2	40
Totals					
4,558	400,378	2,793	1,117	138	6,669
Average					
117	10,266	72	29	4	171
Range	2,204–33,401				

* Half blank. † Fourth blank. ‡ Mixed size paper. § Typewritten. All notebook pages are size 8 × 10½ except the last four in this table which are 8½ × 11, 6¾ × 8½, 7¾ × 9¾, and 5½ × 8½, respectively.

TABLE 5B

PRINTED TYPE NOTEBOOK. NUMERICAL DATA CONCERNING PERCENTAGES OF DRAWINGS AND WRITTEN MATTER, AND VOCABULARIES

Total Number of Pupil Words	Number of Printed Words by Author	Per Cent of Written Matter	Per Cent of Drawing	Number of Pages in Notebook	Name of Author
5,035	26,950	70	30	100	Clement
4,485	26,950	69	31	100	Clement
5,344	26,950	64	36	100	Clement
4,901	26,950	81	19	100	Clement
8,505	26,950	62	38	100	Clement
2,716	26,950	63	37	100	Clement
3,725	26,950	72	28	100	Clement
7,444	7,986	60	40	118	Iroquois
6,104	2,860	58	42	130	Local
17,405	5,215	79	21	156	Local
7,076	420	71	29	84	Local
7,414	17,632	80	20	152	Blaisdell
8,294	17,632	66	34	152	Blaisdell
Totals 88,448		895	405		
Average 6,804		69	31		
Range 2,716–17,405					

TABLE 6

SUMMARY OF DATA IN TABLES 5A AND 5B

Total of Pupil Words, Written and Printed Type Notebooks	488,826
Average Number of Pupil Words per Notebook	9,401
Totals of the Percentages of Written Matter	3,688
Totals of the Percentages of Drawings	1,522
Average Percentage of Written Matter per Notebook	71%
Average Percentage of Drawings per Notebook	29%

and certain numerical figures with reference to vocabulary studies for the written type. Table 5B gives similar data for the printed type. Table 6 gives a summary for both types of notebooks.

The data on the printed type of notebooks in Table 5B are listed under slightly different table headings in order to include the name of the author of the notebook and an estimate of the number of printed words.

An examination of Tables 5 and 6 will show that the relative amount of space in the notebooks given to written matter and drawings varies quite as much in the several notebooks as it varies from the general average of 71 and 29 per cent, respectively. Two of the notebooks have the exact average proportion of 71 to 29, and twelve others are quite near this proportion. The wide range in total number of words used implies a number of things; but pupils write fewer words with the printed type notebook because so much of the work is already presented by the author's printed directions and outlines of the exercises. Variations in the number of words used in the same type of notebook, while to some extent due to individual pupil differences, is largely indicative of lack of technique in notebook writing.

Health Material in the Notebooks

Caldwell and Finley [1] classified biological articles found in newspapers to determine what biological knowledge was of the most value. Among other things, they found that health was stressed more than any other topic. Anita D. Laton [2] in her thesis on "The Psychology of Learning Applied to Health Education Through Biology" found that the diseases of greatest value for study by ninth grade pupils in New York are:

Chicken pox	Colds	Diphtheria	Influenza
Malaria	Measles	Mumps	Pneumonia
Rabies	Scarlet fever	Smallpox	Tuberculosis
Typhoid fever	Whooping cough	Yellow fever	

Her conclusion, in part, is:

> Selection of specific material for teaching, and guidance of children's interests, through application of the psychology of learning can increase the health education value of biology, without increasing the emphasis now being given to such material in textbooks and courses of study.

Claude S. Chappelear,[3] in a study of "Health Subject Matter in Natural Sciences," made in 1927, found the health material in five well known elementary textbooks of biology to average 36.99 per cent of the total content. He also studied thirteen College Entrance examinations and twenty New York Regents

[1] C. W. Finley and Otis W. Caldwell, *Biology in the Public Press.* Lincoln School, Columbia University, 1923.

[2] Anita D. Laton, *The Psychology of Learning Applied to Health Education Through Biology.* Bureau of Publications, Teachers College, Columbia University, 1929.

[3] Claude S. Chappelear, *Health Subject Matter in Natural Sciences.* Ruth Strang, *Subject Matter in Health Education.* Lois Meier, *Health Material in Science Textbooks.* All published by Bureau of Publications, Teachers College, Columbia University.

examinations and found these examinations to contain 33.23 and 31.93 per cent health material, respectively. Similar investigations of the health content of textbooks in other fields have been made, notably the studies in health by Ruth Strang and by Lois Meier.[3] No one, to our knowledge, has made any complete and scientific investigation of the health content in the biological laboratory notebooks of secondary schools.

It was not feasible to make a page analysis of the health content of the notebooks submitted for this study, but by taking the units of written work and applying the material criteria developed by Strang, Chappelear, and Meier, it was found that of the 2,258 units of written work in the fifty-two notebooks examined, 415 units were largely health experiments, exercises, and studies. Therefore, about one-fifth of the notebook material is health matter, as is shown in Table 7.

TABLE 7

Health Content in the Fifty-two Notebooks, Omitting Units of Written Work on Bird Charts, Biographical Sketches, etc.

Classification	Number of Units of Written Work		Per Cent of Health Material Based on Number of Units of Written Work
	Health	Non-Health	
Chemical and Food	155	368	29.63
Zoology	3	265	1.11
Botany	23	969	2.32
Human Physiology	234	2	99.15
Miscellaneous	0	239	0.00
Totals	415	1,843	
Percentage of Health Material			18.38

The percentage of health material found in the notebooks is less than the percentage of health material in the textbooks. This seems to indicate that health study by the laboratory method is not so extensive as it should be. One thing that displaces health study in the laboratory is quite evident—the emphasis on the study of plants. There are 991 units of written work on plants alone in a total of 2,258.

Several notebooks have records of chest expansion of the individual pupils of the biology classes. Many notebooks have the pulse rate exercise. Practically all have material on food, exercise, and physiological functions of the human body. Most notebooks contain some material on the prevention of human disease, sanitation, and hygiene, but no comprehensive study of the diseases found to be of greatest value for study by Laton.[4] No notebook contains any direct reference to social hygiene or euthenics. The subject of eugenics is avoided.

What the Photostats Show

Photostats are presented on pages 25 to 36 to show samples of handwriting; specimens of drawings; and types of written work found in the notebooks.

Photostat No. 10 very clearly illustrates what is meant by units of written work to test for something. No. 8 is a poor example of the chemical type of experiment. No. 11, marked "Dictated," illustrates dictated material and it also illustrates the report type of writing up an exercise. No. 9 shows a unit of written work on the human phase of biology and serves as an illustration of the usual way of reporting laboratory work. Photostats 1 to 7 inclusive show specimens of drawings selected from six different notebooks. Their special significance is discussed in Chapter III in connection with the study made of drawings in all the notebooks.

The selection of the photostatic material from the notebooks was not made specifically to show quality of handwriting so much as to show types of units of written work and specimens of drawing, but as a random sampling of the quality of handwriting in the notebooks the examples may be taken as fairly representative. Handwriting in the notebooks is discussed in Chapter IV.

Material in the Notebooks of Value in Prevocational Training and Training for the Worthy Use of Leisure

The units of written work are nearly all indirectly suggestive of occupational interests if the teacher had vocational guidance in mind when teaching biology. The study of human food is found in most of the notebooks, but such occupations as cook, chef, dietetian, etc., are not mentioned. Most of the notebooks have

[4] *Op. cit.*

units of work on the conservation of forests, which should suggest such occupations as forester, nurseryman, and forest ranger, and probably it does to some pupils. Every notebook has some work reported on bird study. There is nothing, however, in any notebook about domestic fowls, chicken hatcheries, and poultry farming. The study of animals, as reported in the notebooks, does not suggest dairying, fur farming, or animal husbandry, but the situation is somewhat different with respect to the study of plants. The study of plants does suggest gardening. Practically every notebook has units of written work on flowers and in several there are studies of fruits, but none of this work is organized in a way that would be suggestive of such occupations as agriculturist, plant breeder, etc.

There is practically nothing in the notebooks that directly suggests training for the worthy use of leisure; but field trips are mentioned, and in one notebook there is reference to a visit to a museum of natural history. Many of the notebooks have references to sportsmen, but always emphasizing that hunters are enemies of wild life, no distinction being made between real sportsmen and the indiscriminate killers of song birds. The study of fish is reported and mention is made of commercial fishing, but trout is the only game fish studied. The use of the bird glass and camera in bird study is not mentioned, either as a method of bird study or as a pleasant way to enjoy leisure.

CHAPTER III

STUDY OF THE DRAWINGS IN THE NOTEBOOKS

The drawings in the notebooks were counted and classified as typical, human physiology, chemical and food, plant, and lower animal, according to the subject matter illustrated. By typical are meant illustrations of a model plant or animal for purposes of laboratory observation and study. Each separately labeled illustration was counted as a separate drawing. Chart I illustrates graphically the relative proportion of drawings with reference to the classification of subject matter in the notebooks.

The drawings were further classified as analytical and non-analytical for each of the subdivisions enumerated. By analytical are meant those drawings which show structures not visible to the naked eye without dissection or the use of the microscope.

This classification shows that the emphasis is on the non-analytical phase for 62 per cent of the plant drawings, 86 per cent of the lower animal, and 30 per cent of the human physiology are of this non-analytical type. The textbooks in general condemn the representative or picture type of drawing, and Ayer [1] finds it uneconomical of time. Yet this practice is based largely on the objectives of the old-type biology. This study tends to show how eager pupils are to add the picture effect to the drawings, and future studies might reveal a worthwhile objective in this phase of the work of modern biology.

There is little uniformity as to what is illustrated, even with the same subject matter.

Table 8 gives a summary of the data evolved from the study of all the drawings in the notebooks.

There are no analytical drawings for lower animals in twelve notebooks; fifteen have none on the human phase; eight have none on chemical and food; and one has no non-analytical drawings for lower animal.

[1] Fred C. Ayer, in *Third Yearbook, Department of Superintendence*, National Educational Association.

CHART I

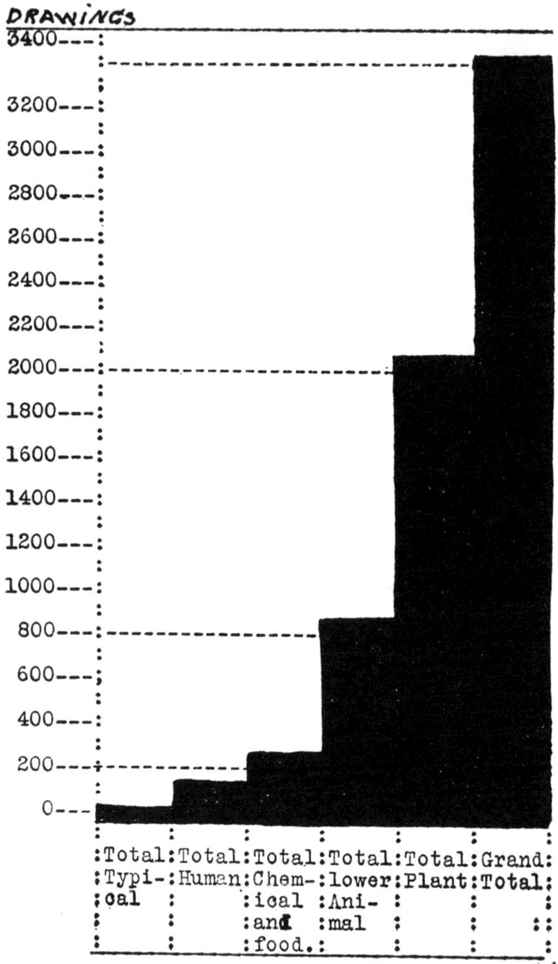

GRAPHIC COMPARISON OF DRAWINGS TO CLASSIFICATIONS OF SUBJECT MATTER

TABLE 8

STATISTICAL SUMMARY OF THE DRAWINGS IN THE FIFTY-TWO NOTEBOOKS

Total Number of Drawings		3,357
Average Number of Drawings per Notebook		64.55
Range of Number of Drawings in the Notebooks		24–154
	Total	Aver. No
Drawings Classified as to Subject Matter:		
Plants	2,048	39.38
Lower animal	851	16.36
Chemical and food	259	4.98
Human physiology	168	3.23
Typical	31	0.59
	Aver. No.	Range
Drawings Classified—Analytical Type:		
Plant	15.05	6–35
Chemical and food	4.98	0–32
Lower animal	2.21	0–15
Human physiology	2.25	0–16
Typical	0.59	0– 7
Colored drawings	2.90	0–33
	Aver. No.	Range
Drawings Classified—Non-Analytical Type:		
Plant	22.32	8–71
Lower animal	13.15	0–45
Chemical and food	0.00	0–00
Human physiology	0.98	0–16

About 85 per cent of the drawings are made with pencil; the remaining 15 per cent are made with ink. Some of the drawings are colored with crayon and a number are shaded. Accuracy and proportion is not a general characteristic of the drawings. Practically all drawings, on the other hand, are correctly labeled. There is no evidence that the drawings were made from actual specimens, and any drawing in any notebook may have been copied from a book. Reference to photostats No. 1 and No. 1a shows what appears to be a clear case of copy work. Photostat No. 1 is reproduced from "The Laboratory Manual" by T. J. Moon. Photostat No. 1a is reproduced from a student's note-book.

Photostat No. 7, marked "Satisfactory," on the original, shows crowding of illustrations on a single page, and the drawings appear to be copied. Photostat No. 5 is an example of a good drawing and one that may not have been copied. Photostat No. 4, "The Human Skeleton," marked "Approved," represents the old type of drawings. Photostat No. 6, "Dorsal View of the

Frog," is a representative drawing which is an undesirable type, but No. 2, "The Necessity of Oxygen for Germination," is an excellent illustration. This is a drawing that shows controlled experimentation and really means much in a notebook. There are few such drawings in the notebooks submitted for this study. Photostat No. 3 shows a drawing without any labels or even a title, taken from one of the notebooks, the original of which was stamped "Approved" by the teacher. Photostat No. 9 shows how a drawing may be placed on the same page with the written matter. There is much to be said in favor of this practice. Photostat No. 8 is an example of drawing taken from a notebook to show the usual practice of illustrating apparatus in chemical experiments.

Most of the drawings have titles, and most of them are drawn on separate pages, especially the more extensive drawings. Practically all the drawings are freehand sketches, but in many cases the drawings are of the picture-sketch type, instead of the diagrammatic. In this regard, Fred C. Ayer [2] says: "The diagrammatic drawing is more economical of time than is the picture sketch, and also, even with less time put upon it, is superior in fixing in the pupil's mind the facts to be retained."

A study of the drawings indicates that probably too much of the pupil's time is devoted to drawing and possibly much of this time is spent in copy work. There is evidence of time wasted in trying to display artistic skill and in unnecessary coloring of some of the drawings. A drawing should be a record of the experiment or exercise rather than the experiment or exercise itself.

[2] Fred C. Ayer, *op. cit.*, p. 300.

PHOTOSTAT NO. 1

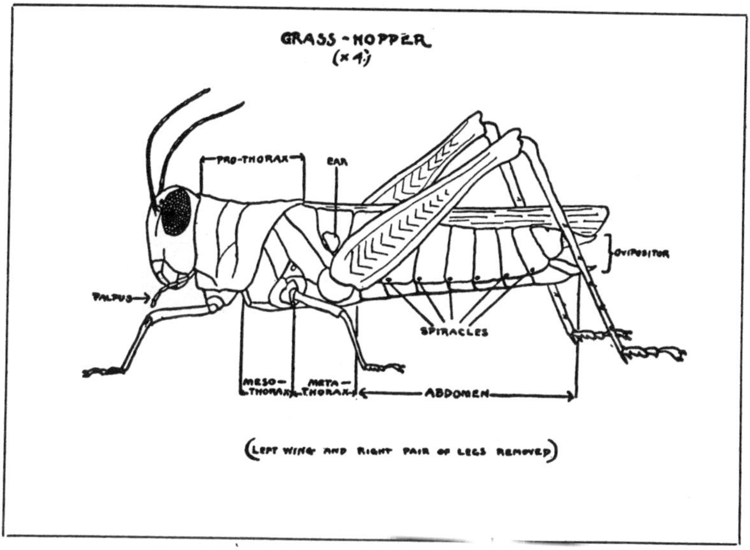

PHOTOSTAT NO. 1A

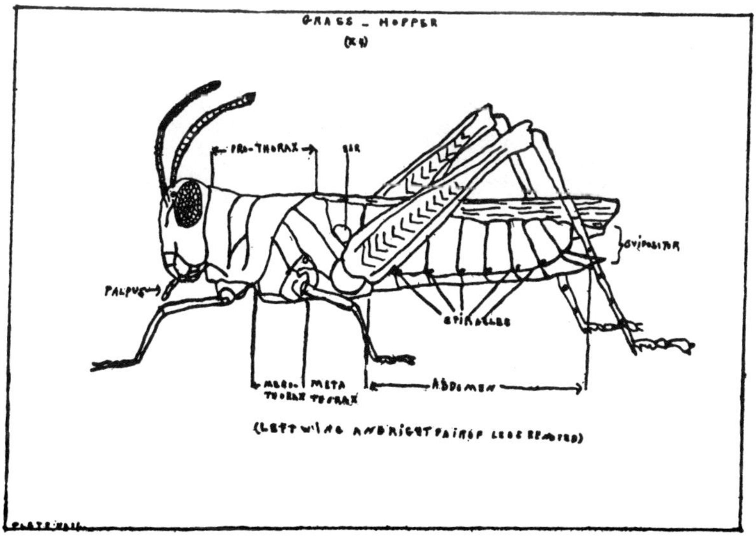

PHOTOSTAT NO. 2

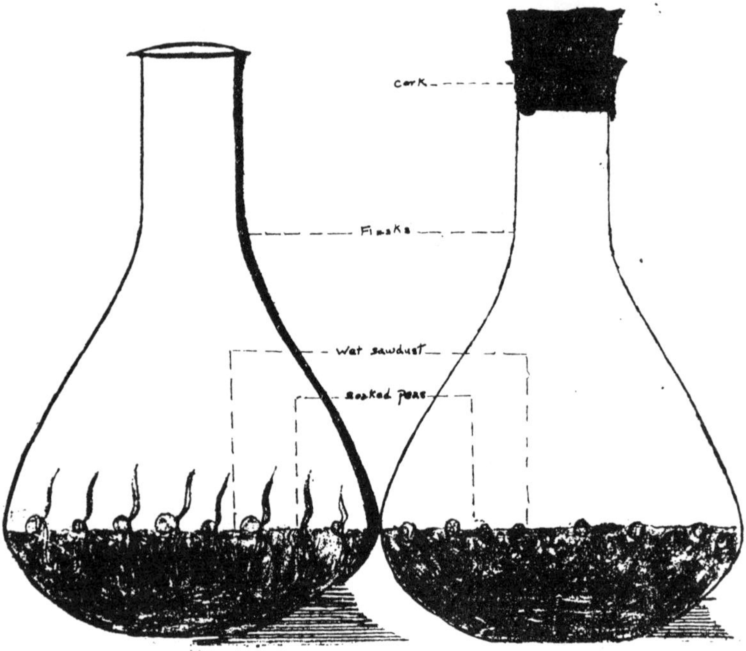

PHOTOSTAT NO. 3

PHOTOSTAT NO. 4

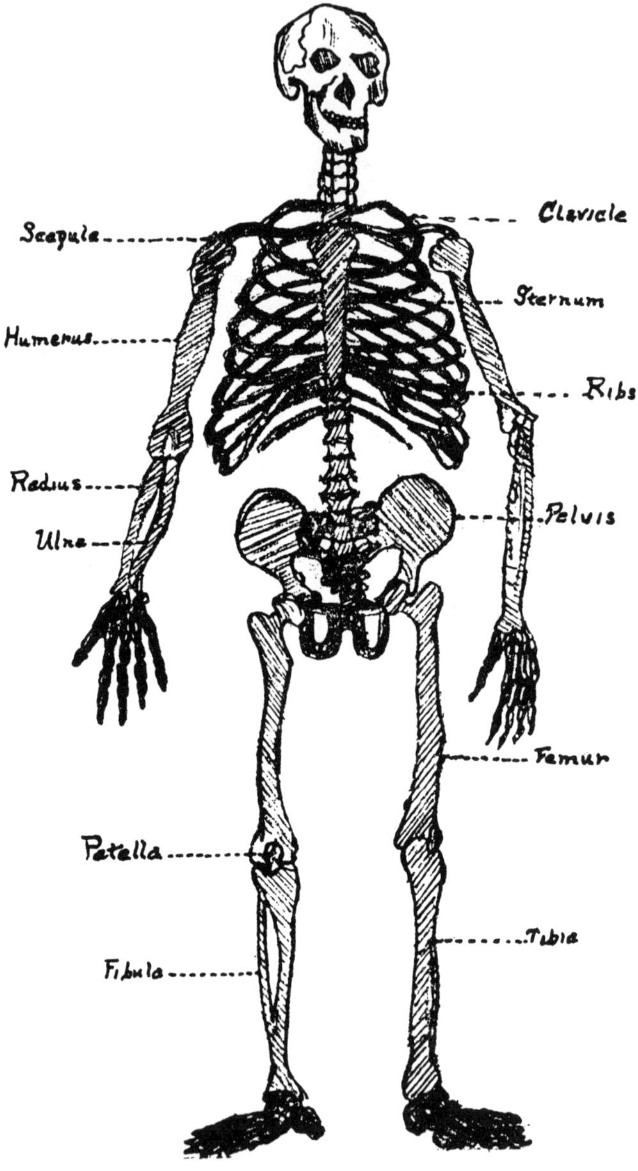

PHOTOSTAT NO. 5

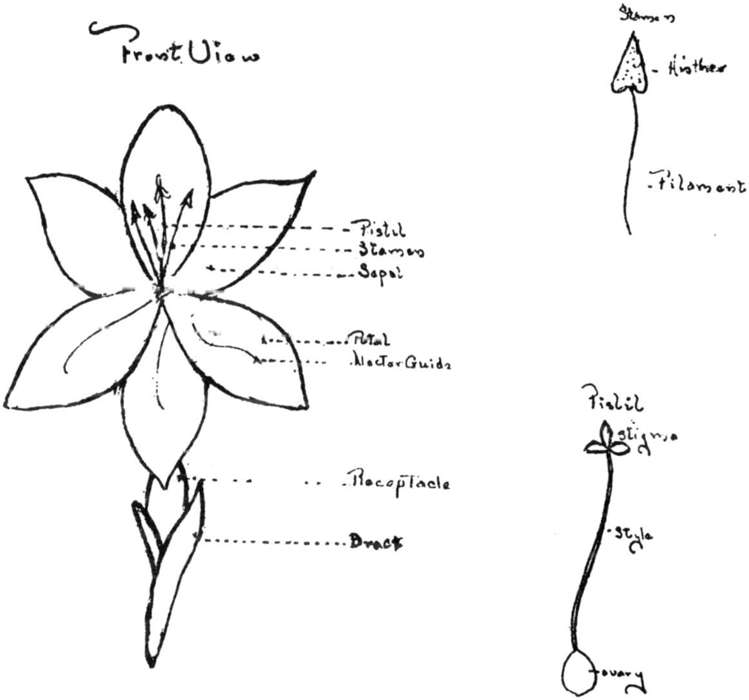

PHOTOSTAT NO. 7

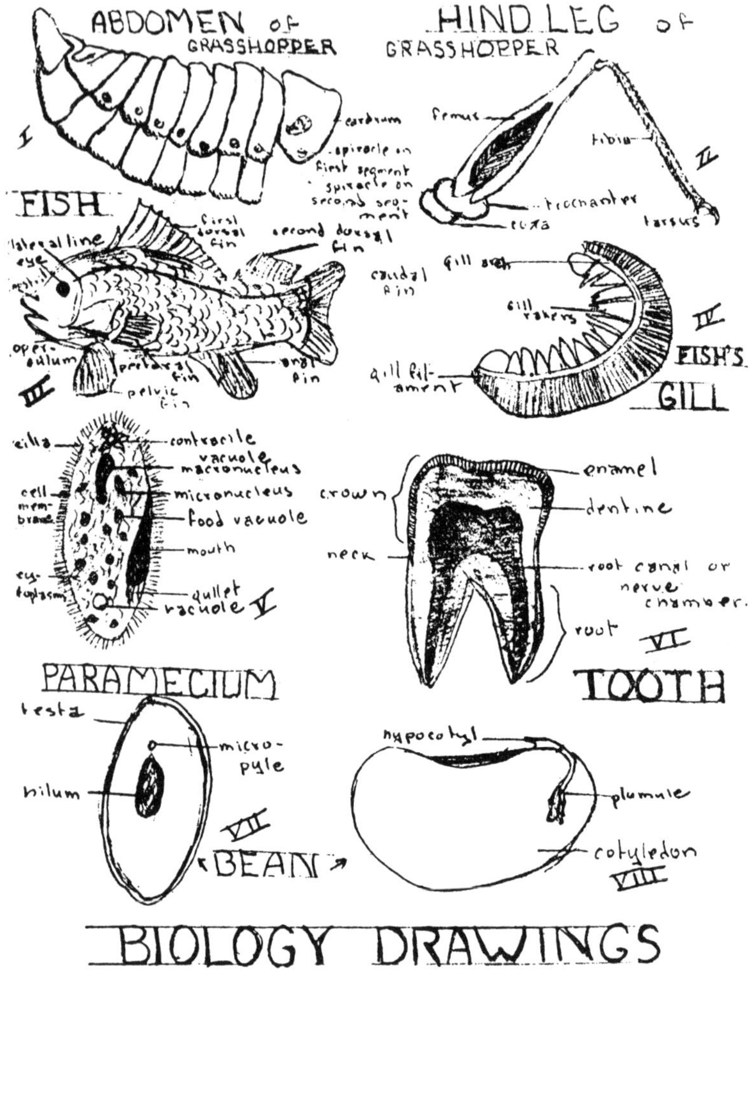

PHOTOSTAT NO. 9

Capullarie Circulation Experiment No. XXIX

Object.
To observe blood as it flows through the capullaries in the web of a frog's foot

Materials.
Live frog, microscope, a wet cloth, a board with a hole in it, and pins.

Method.
We wrapped the frog in a wet cloth and fastened it to the board. We stretched the membrane in the toes over the hole in the board and pinned them We next placed the board under the microscope.

Observation.
We saw the blood flowing through the capullaries, we could see the corpuscles inside the blood as the walls of the capullaries are very thin

Conclusion.
See the drawing below.

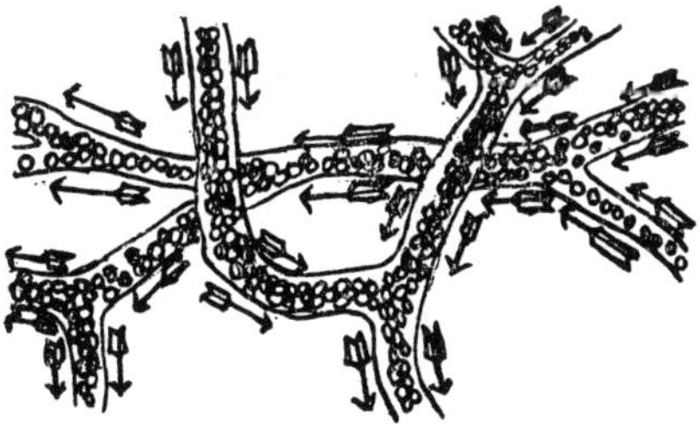

PHOTOSTAT NO. 10

Sugar

Object: To test for presence of sugar in food.

Note: A substance containing sugar when heated with Fehling's solution. turns brick red.

Materials: candy. glass tube. Fehling's solution. heater.

Procedure. The candy is placed in the glass tube. and Fehling's solution added to it. Then the tube is placed over the fire and heated.

Observation: It is observed that the candy turns brick red.

Conclusion: I conclude that there is sugar present in the candy.

PHOTOSTAT NO. 11

Oxygen Cycle

Oxygen is taken in from the air or the water by plants or animals & takes part in oxidation. A waste product called CO_2 results. CO_2 is taken up by oxygen plants & the carbon becomes part of starch while the oxygen escapes into the air again to be used again by some plant or animal in respiration. Oxygen can go thru this series of changes of times without losing any of the properties. It is thus said to go in a cycle or circle in its relation to living things

PHOTOSTAT NO. 6

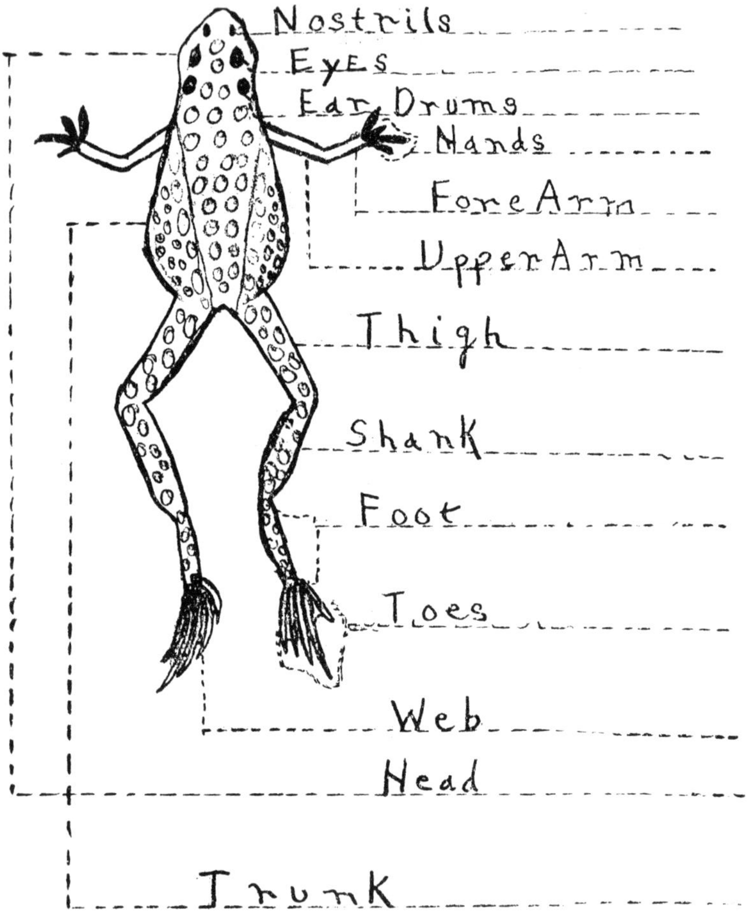

PHOTOSTAT NO. 8

Acr

Object : to make + study oxygen

Method

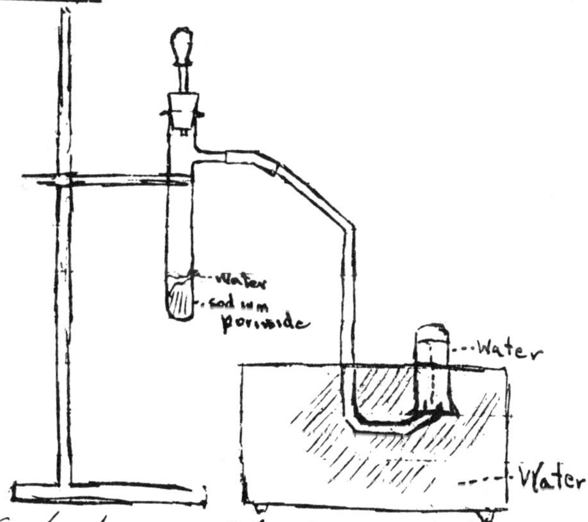

We put in the delivery tube some soidium paroxide put a drop of water on it from the dropper

Observation I saw when the water hit the soedium paroxide that it bubbled and bubbles came up through the tube into the bottle The tube was hot

PHOTOSTAT NO. 8 (*Continued*)

Exp. II

Method

We took the glass + put glass plate over it then we took a glowing splinter and it glowed flamed up

Observation I saw that the glowing splinter burn

Chemests say Oxygen supports combustion

III

Object To see what is now in the bottle

method: we put some clear lime water in the bottle and shook it up

Observation. The lime water turned milky.

Conclussion. Carbon dioxide was in the battle because carbon dioxide is the only thing that will make lime water milky

CHAPTER IV

THE VOCABULARY AND HANDWRITING STUDIES

The Vocabulary Study

The pupil vocabulary in the fifty-two notebooks was examined to determine the following: extent of pupil vocabulary; extent of the use of scientific terms; and correct use and spelling of words. A summary of the notebook vocabulary is given in Tables 9 and 10. A list of the words not found in the Thorndike *Teacher's Word Book* or the Powers *List of Scientific Terms for High School Students* is given in Table 11.

The total of all the separate words used in the notebooks was checked against the Thorndike list and the Powers list. In making up the list of words used in the notebooks proper names and symbols were counted in the total count for all the notebooks. The first step in discovering the extent of the pupil vocabulary as represented in the notebooks was a study of one of the smaller and one of the larger notebooks for comparison and to provide an initial word list for further study. Table 9 shows the results from the comparative study of the two notebooks.

TABLE 9

Summary of Vocabulary Study of One of the Smaller and One of the Larger Notebooks

	Smaller Notebook	Larger Notebook
Number of Words Used but Once	167	653
Number of Words Used 1 to 4 Times	313	1,223
Number of Words Used 5 or More Times	109	558
Total Separate Words Used	422	1,781
Number of Proper Names	15	121
Number of Symbols Used	10	74
Number of Abbreviations Used	4	4
Total Number of Words Used	2,220	15,249
Total Number of Words, Including Symbols, Proper Names, and Abbreviations	2,259	15,448
Number of Words Not Found in Thorndike's List	61	287
Number of Scientific Terms Not Found in Powers' List	7	68

CHART II

GRAPHIC COMPARISON OF FREQUENCY OF SEPARATE WORDS, ALPHABETICALLY ARRANGED, IN ONE OF THE SMALLER AND IN ONE OF THE LARGER NOTEBOOKS

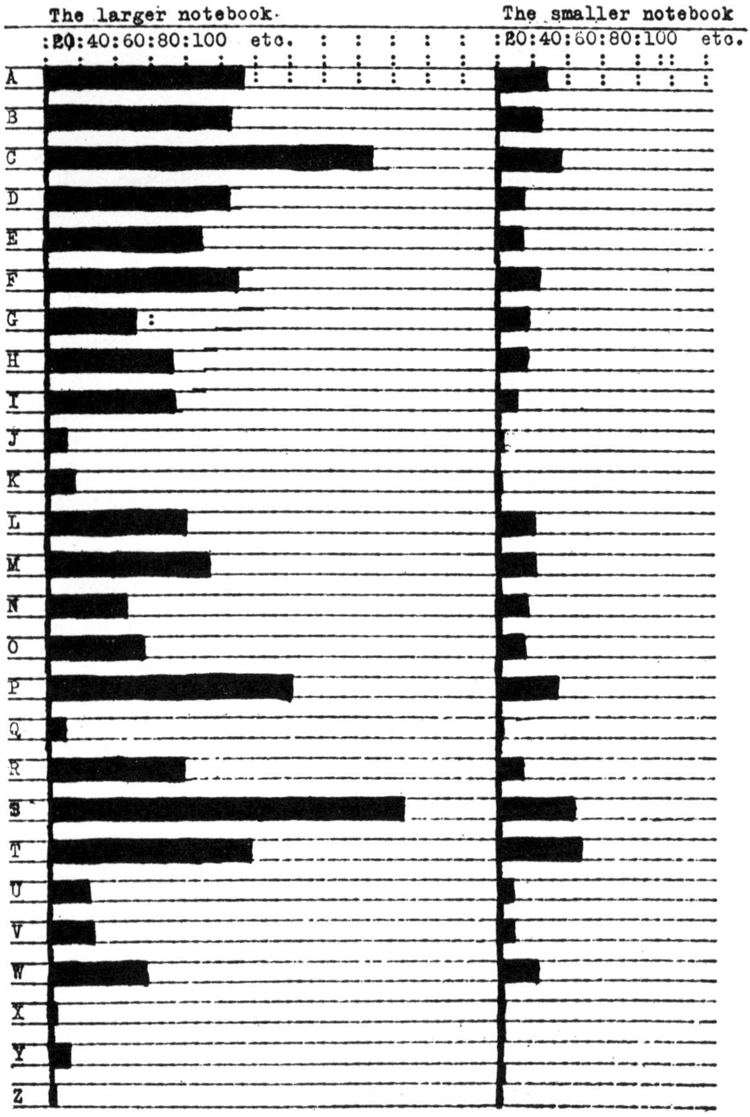

Chart II gives a graphic comparison of the frequency of the separate words, alphabetically arranged, in one of the smaller and in one of the larger notebooks.

The total number of scientific terms used in all the notebooks is 645. Many of the scientific terms were used but once in a single notebook. Comparatively few words were found to be misspelled and but few words "coined" by the pupils. Table 10 gives a summary of the vocabulary study of all the notebooks.

TABLE 10

SUMMARY OF THE VOCABULARY STUDY OF ALL THE NOTEBOOKS

Total Number of Words Used	488,826
Total Number of Separate Words Used	2,358
Total Number of Words Not Found in Thorndike's List	344
Total Number of Words Not Found in Powers' List	87

Only one typewritten notebook was submitted for this study, but it is not unlikely that future high school students will submit typewritten notebooks to the science teacher. Class notes and notes on laboratory work must, however, continue to be written by hand. Therefore it seems that some effort to secure better handwriting is desirable.

TABLE 11

WORDS IN THE NOTEBOOKS NOT FOUND IN THORNDIKE'S TEACHER'S WORD BOOK, OR IN POWERS' LIST OF SCIENTIFIC TERMS FOR HIGH SCHOOL STUDENTS

Words starred not in Powers' list

Absorption	Appendage	Beneficial
Adaptation	*Arterial	Bichloride
Adenoids	Artery	*Biconcave
*Advocacy	*Ashy	*Bicuspid
Aeration	Assimilate	Bile
Agar agar	*Assimilation	Biology
*Ahead	*Astigmatism	*Blotter
Alimentary	Auditory	Boll
Amoeba	Auricle	*Box elder
Amphibia		*Bronchial
*Anal	Bacillus	Burdock
Anopheles	Bacteria	
Antennæ	Bacteriology	Calyx
Antibody	*Bag-like	Cambrium
Anther	*Bailers	Capillary
Aorta	Bast	Capsule
*Apex	*Batting	Carapace
Aphid	Beaker	Carbohydrate

TABLE 11—*Continued*

Carpel
*Carrion
Caudal
Cavity
Cellulose
Cephalothorax
Cerebellum
Cerebrum
Char
*Chitin
*Chlorate
Chlorophyll
Chloroplast
*Chyle
*Chyme
Cilia
Ciliate
*Cinnamon
Cloaca
Clot
*Coccus
Cochlea
*Cochineal
Cocklebur
Codlin
Combustion
Complex
Concave
Conservation
Convex
*Convolution
Coordinate
*Coordination
Corolla
Cornea
Corpuscle
Cortex
Cotyledon
*Coxa
Crayfish
*Cross bill
Cross-pollinate
*Cross section
Crucible
Crustacea
Cytoplasm

*Daddy longlegs
*Decrease
Dehiscent
*Derangement
Devastate
*Dexterity
Diastase
*Dissect
*Dissension
Dicotyledon
Dioxide
Disinfectant
Dormant
Dorsal
Drupe

*Ectoplasm
Edible
Embryo
*Emulsion
Encyst
Endosperm
*Endoplasm
Enzyme
*Epicotyl
Epidermis
Erosion
Esophagus
Ether
Eustachian
*Everytime
Excretion
Exoskeleton
Explode

Facet
Fahrenheit
*Fan-like
Fehling
Fermentation
Filament
Filter
*Flower-like
Focus
Follicle
*Fork-like
Formaldehyde
Fumigate

Gastric
Generalization
Generate
Germinate
*Glottis
Gram
Grippe
Gullet

Heliotropism
*Hessian
*Hibernation
Hydrochloric
Hydrogen
Hydrophobia
Hydroxide
Hygiene
Hygienic
Hypocotyl
Hypo-pharynx

Ichenumon
Ignite
Indigestible
Inorganic
Insecta
Insoluble
Integument
Katydid

Labrum
Lacteal
Larva
Larynx
Lateral
Legume
Lengthwise
Lenticel
Liberate
*Licorice
Limestone
*Lysol

Maggot
Malaria
Mammal
Mandible

TABLE 11—*Continued*

*Mandrake
Masticate
Maxilla
Medulla
Medullary
Membrane
Membranous
Merge
Mesothorax
Metamorphosis
*Metathorax
Metazoan
Micropyle
*Microsca
Microscope
Microscopic
*Mold
Mollusk
Monocotyledon
Monocotyledonous

Narcotic
Necessitate
*Neuron
Nicotine
Nitric (acid)
Nitrogen
Nitrogenous
Nodule
*Non-fatty
Nucleus
Nutrient

*One-celled
Operculum
Opium
Osmosis
Ovary
*Over done
Ovipositor
Ovule
Oxidation
Oxidize
Oxygen

Palisade
Palmate
Palpus
Pancreas
Pancreatic
Papilla
Paramecium
Parsnip
Pectoral
Pelvic
Pelvis
*Pericardium
Petiole
Pharynx
Photosynthesis
Physiologic
Pigment
Pinnate
*Pisces
Pistil
Placenta
Plasma
*Pliable
Plumule
Pollination
Pome
Posterior
Potassium
Protein
*Prothorax
Protoplasm
Protozoa
Pseudopodia
Ptyalin
Pulmonary

Quarantine

Rabies
*Rakers (Gill)
*Raphe
Reflex
Reproduce
Reproduction
*Respiration
Respiratory
Retina
Rhubarb
Rodent
*Rod-shape
Rootlet

Saliva
Salivary
*Salts
*Sawdust
Scavenger
Scientist
*Sclerotic
Sebaceous
Secrete
Secretion
Segment
Segmental
Self-control
*Semicircular
*Semifluid
Sensory
*Sheeting
Soluble
Spiracle
Spirillum
Spore
*Sputter
Sputum
Stamen
*Starling
*Steapsin
*Stegomyia
Sterilization
Stigma
Stimulate
Stimulant
Stimulus
*Stipule
Stoma
Stomata
*Stopper
Suffocate
*Sunlight
Swimmeret

Tachina fly
Tactile
*Taproot
*Telson

TABLE 11—*Continued*

Testa	*Trillum	Vaccination
Thoracic	*Trochanter	Vacuole
Thorax	Trypsin	Ventral
Tibia	*Tsetse	Vertebrate
Trachea	*Tweezers	Vireo
Translucent	*Typhoid	
*Transparent		Weevil
Transpiration	*Unglaze	
*Trichocyst	Urea	Xylem
*Tricuspid	Ureter	

The Handwriting Study

The handwriting of pupils submitting notebooks, except in two notebooks that were hand printed and one that was typewritten, was measured with the Ayres Handwriting Scale, Gettysburg edition. The results of this measurement of the quality of the handwriting in the notebooks are tabulated in Table 12.

TABLE 12

Handwriting Scores of the Pupils Submitting Notebooks as Found in Measuring with the Ayres Handwriting Scale, Gettysburg Edition

Number of Pupils	Score	Totals
3	20	60
7	30	210
3	40	120
10	50	500
8	60	480
7	70	490
4	80	320
7	90	630
Totals 49*		2,810
Average Score		57.35

* One notebook was typewritten and two were hand printed.

Several noteworthy studies of handwriting have been made to determine the quality desirable for social and business usage. Leonard V. Koos,[1] of the University of Minnesota, in his study

[1] Leonard V. Koos, "The Determination of Ultimate Standards of Quality in Handwriting for the Public Schools." *Third Yearbook, Department of Superintendence*, National Education Association, 1925. pp. 209-11.

of handwriting, secured the judgment of 826 adults, including school teachers, to determine the quality of handwriting necessary for social correspondence. The majority of the people whom he selected as judges considered a quality of 60 on the Ayres Handwriting Scale satisfactory. Professor Koos also determined, by measuring the handwriting of 1,127 employees in a variety of occupations and having these specimens of handwriting judged by a number of persons, that a quality of 60 on the Ayres Scale is adequate for most purposes of industry.

Professor Frank N. Freeman,[2] of the University of Chicago, in a study of handwriting sent a questionnaire to fourteen business firms to find out the importance of handwriting and to determine the quality necessary in business. The prevailing opinion was that 60 on the Ayres Scale is satisfactory.

Using Ayres Handwriting Scale, Gettysburg edition, the University of Iowa, Extension Division, in 1927 conducted a statewide investigation of handwriting in the public schools. The norms for Grades 9 to 12 were not established, but an average of the composites for Grades 9, 10, 11, and 12 is approximately 42½ in quality on the Ayres Scale.

The average of the scores for the handwriting in the notebooks submitted for this study, as shown in Table 12, is 57.35 on the Ayres Scale, and is therefore fairly close to the quality thought desirable for social and business purposes. The range in quality, however, is wide. Reference to Photostats Nos. 8 to 11 will give a fair idea of the quality of handwriting found in the notebooks.

[2] Frank N. Freeman, "Research in Constructing the Elementary School Curriculum." *Third Yearbook, Department of Superintendence,* National Education Association, 1925. pp. 206-9.

CHAPTER V

THE RELATED FACTORS

Textbooks

Although but six different books were reported as texts in biology in the cooperating schools, a number of other tests were reported as being used as supplementary or reference texts. In this study the ten textbooks selected for analysis, include the six used in the cooperating schools and four from the supplementary list. The data given in the tables represent the findings for the six official textbooks in biology as reported by the cooperating schools; Chart III gives data for the textbooks selected for analysis.

In the analysis of the biology textbooks the mechanical features were assumed to consist of the number of pages of supplementary matter, visual and material, material in the text proper, and total number of pages of all materials in the textbook.

The number of pages of supplementary matter was further classified as: preface, index, glossary, summary, review questions, list of illustrations, collateral readings, and title pages.

The visual aid material was classified as: maps, charts, graphs, drawings, and photographs. Everything not a map, chart, graph, or photograph was listed as a drawing. The photographs were listed under the headings: microscopic, civic and health appeal, still-life, and general. This classification of the photographs is more or less arbitrary. The placing of a particular reproduced photograph under a particular heading was largely based on personal opinion; one person, for example, may see in a particular photograph a civic urge or health appeal that another would consider non-existent. Any picture taken by means of a microscope and camera is listed as a microscopic type photograph. Under the heading of still-life photographs were listed all pictures taken of animals in their natural environment.

Material in the text proper includes botany, zoology, human physiology and hygiene, and biology. The term biology in this connection means the treatment of plants and animals as related units in life. The number of pages for each division or subject was obtained by measuring, with a rule, the number of inches of printed matter devoted exclusively to botany, zoology, etc. The total number of inches for each subject was then translated into terms of pages. If an average page of a textbook contained seven inches, vertically measured, of printed matter and there was a total of 77 inches of botany in the book, the number of pages devoted to botany would be 11.

There is very little agreement on the part of the authors of the textbooks in biology concerning the mechanical features of the texts. The number of chapters varies considerably and the amount of supplementary material in ten textbooks ranges from 53 to 133 pages. Some texts have a glossary and others do not. There is some disagreement, as indicated in the "Aims and Objectives" of the authors, concerning the inclusion of laboratory exercises in the textbook.

(See Tables 13, 14, 15, 16, and 17, and also Chart III, on page 51, for the record of the textbook analysis.)

Analysis of the Textbooks of Biology Used in the Co-operating Schools

Aims

The aims and objectives of the biology course formulated by the authors of textbooks for biology as stated in the Introductions to their books follow:

"It is the conviction of the author that a first-year high school course in biologic science should be based on the conception of biology as a single science rather than as three entirely separate courses in zoology, physiology and botany. It should be one of the purposes of a biology course to teach the great ideas or principles that underly all life. The best way to assure this is to ascertain as soon as possible what these principles are and then to explain their workings as applied to animals, plants and man."—Arthur G. Clement, *Living Things,* pp. iii, Preface.

". . . The culmination of such an elementary course is avowedly the understanding of man, and the principles which hold together such a course should be chiefly physiological."—George W. Hunter, *New Essentials of Biology,* pp. v, Preface.

"While the place of biological science is not fixed in all parts of this country, the tendency is well marked to place it in the tenth year of the school curriculum. Recognizing this, *New Civic Biology* has attempted to build on the unorganized science facts that the high school pupil has already assimilated. The trend toward better health and citizenship building has been recognized, and it is hoped that this book will work toward the ideal development of efficient, thinking citizens."—George W. Hunter, *New Civic Biology*, pp. vii, Preface.

". . . The course here presented emphasizes the fact that biology is a unit science, based on the fundamental idea of development, rather than a forced combination of portions of botany, zoology, and hygiene."—Thurman J. Moon, *Biology for Beginners*, pp. v, Preface.

"Twenty years ago, the authors, in common with many other earnest teachers of biology, became convinced that much more emphasis should be placed on the functions of living organisms as contrasted with their structure and that a comparative study should be made of plants, animals, and the human body . . . *Biology and Human Welfare* represents the development of our ideas, based on the observation and experience of ten years and more since the publication of the earlier book, of the best means of realizing these aims. . . . The dominating purpose throughout this book is to show the intimate relation of biological science to human welfare. . . ."—Peabody and Hunt, *Biology and Human Welfare*, pp. v-vi, Preface.

"*The New Biology* offers a unit course dealing with the fundamental functions of living things. . . . We are coming to realize more and more that it is what an organism does that is important rather than the details of its structure, and this is the reason that we have departed somewhat from the older type study which has characterized earlier texts on biology."—Smallwood, Reveley, and Bailey, *New Biology*, pp. iii, Preface.

Directions

The following are typical of the directions for laboratory and notebook work included by authors of textbooks of biology:

"As far as possible, observation of animals and plants and exercises involving experiments should precede the study of the text. Although general outlines for laboratory work are given, elaborate directions for these exercises have been omitted since each teacher usually prefers to present these in his own way, varying them according to the facilities available, in different schools. The subject matter may often be presented advantageously in the form of projects . . ."—Arthur G. Clement, *Living Things*, pp. v, Preface.

"The laboratory exercise immediately precedes the textbook discussion, the latter being used to clear up any false inferences the pupil may have made from the specimen in hand and to fix the object of the problem in the mind of the pupil. Too often has a laboratory exercise meant nothing to a pupil but 'busy work'. A plainly outlined and organized plan of attack, a few references to the text or to previous work performed, and a definite problem will result in better and more definite laboratory work. For use with this book manuals for the solution of laboratory problems have been prepared . . . . . . The latter manual, *'Laboratory Problems in Civic Biology'*, gives an excellent point of attack through the interests of the student and gives a varied selection of problems for the teacher."—George W. Hunter, *New Essentials of Biology,* p. vii, Preface.

". . . . Ideas, not types, should be the ultimate development of the laboratory work . . . Following the problems (in the text) are laboratory suggestions, many of which are worked out fully in *Hunter's Laboratory Problems in Civic Biology.*"—George W. Hunter, *New Civic Biology,* pp. vii-viii, Preface.

"Outlines, tabulations, diagrams, and vocabularies take up a larger proportion of pages than in any similar text. The numerous diagrammatic line drawings are intended to simplify matters of structure for the beginner who would have difficulty in selecting the essential points of a photograph or highly detailed line drawing. It is hoped also that a reasonable use of line drawings will help the pupil in his laboratory work by affording models which he can easily approximate. . . . No laboratory work is included in the text. Such a plan tends to invite copying from the book rather than gaining the information from observation of the material in hand. In a separate manual such as the author has prepared more complete directions can be given and more detailed results required."—Thurman J. Moon, *Biology for Beginners,* pp. v-vi, Preface.

"Laboratory work is most important in arousing the interest of pupils and in fixing in their minds essential facts and principles. We have chosen for laboratory study plants and animals that are available in most parts of the country. The laboratory exercises have been included in the body of the book, because they constitute an essential part of the course and because they are more convenient for use here than they would be if placed in a separate volume or in the appendixes."—Peabody and Hunt, *Biology and Human Welfare,* pp. vii, Preface.

"Biology will take its recognized place as a valuable subject in training only when teachers require accuracy of observation and the recording of these observed facts in suitable notebooks. It is possible to learn much about our own lives from the study of animals and plants, but trustworthy conclusions can be formulated only from reliable observations. We have suggested a method of study in the laboratory problems to train the student in accuracy of observing and recording."—Smallwood, Reveley, and Bailey, *New Biology,* pp. iii, Preface.

Table 13 shows that in the six textbooks examined there is lack of uniformity in the division of the subject matter into chapters and quite a significant variation in the amount and classification of supplementary material.

TABLE 13

MECHANICAL FEATURES OF THE BIOLOGY TEXTBOOKS

Key to Author	Number of Pages of Supplementary Matter*	Number of Pages of Visual Aid Material	Total Number of Pages	Number of Chapters	Number of Pages of Summary	Number of Pages of Questions
A.**	86	61	viii + 488	45	38½	12½
B.	74	51	xii + 448	33	0	27½
C.	55	61	x + 453	30	7	7½
D.	131	71	xxv + 647	57	93	0
E.	72	76	xii + 585	27	19	15½
F.	129	110	xix + 741	37	37½	10½
Totals	547	430	lxxxvi + 3,362	229	195	73½
Average ..	91.16	71.66	xiv + 560.33	38.16	32.50	12.25

* In listing the total number of pages in each textbook, the number of extra pages numbered viii, xii, etc., does not include all the supplementary material in a given text.

**Authors	Name of Textbook	Copyright Date
A. Clement	*Living Things*	1924
B. Hunter	*New Civic Biology*	1926
C. Hunter	*New Essentials of Biology*	1923
D. Moon	*Biology for Beginners*	1926
E. Peabody and Hunt	*Biology and Human Welfare*	1924
F. Smallwood, Reveley, and Bailey	*New Biology*	1924

Table 14 shows how the authors regard etchings and drawings; the range is between 114 and 200 drawings for each biology textbook.

Tables 15 and 16 show the proportion of supplementary and text material in biology textbooks. The greatest divergence is observed in the number of pages of references. The range is from 2 to 12 pages. The number of pages of biology varies from 16 to 253.

Table 17 indicates that marked emphasis is placed on health material.

TABLE 14. VISUAL AID MATERIAL IN THE TEXTBOOKS

Key to Author	Number of Maps	Number of Graphs	Number of Charts and Tables	Number of Etchings or Drawings	Reproduced Photographs: Microscopic	Civic and Health Appeal	Still Life	General	Total
Clement	0	0	7	114	2	26	10	43	202
Hunter*	7	21	27	188	9	21	5	54	332
Hunter†	5	12	7	189	12	8	13	110	356
Moon	2	3	28	118	2	3	13	41	210
Peabody and Hunt	1	4	6	200	9	24	19	93	356
Smallwood, Reveley, and Bailey	1	6	7	185	8	10	51	170	438
Totals	16	46	82	994	42	92	111	511	1,894
Average	2.66	7.66	13.66	165.66	7	15.33	18.5	85.16	315.66

* Hunter's *Civic Biology*. † Hunter's *New Essentials of Biology*.

TABLE 15. SUPPLEMENTARY MATERIAL IN THE TEXTBOOKS

Key to Author	Number of Title Pages	Number of Pages in Preface	Number of Pages in Table of Contents	Number of Pages for Laboratory and Field Work	Number of Pages of Reference and Collateral Readings, Including Bibliography	Number of Pages in Glossary	Number of Pages in Index
Clement	3	3¼	2	0	2	11	16
Hunter*	2	3	2	7½	7	14	18
Hunter†	2	4	1½	0	7	13	13
Moon	3	4¾	2	0	12	0	16¼
Peabody and Hunt	2	5	1¼	13	3	0	12¼
Smallwood, Reveley, and Bailey	2	2	15	0	3	33½	36½
Totals	14	22	23¾	20½	34	71½	112
Average	2.33	3.66	3.96	3.42	5.66	11.92	18.66

* Hunter's *Civic Biology*. † Hunter's *New Essentials of Biology*.

TABLE 16

TEXT MATERIAL IN THE TEXTBOOKS

Key to Author	Number of Pages of Botany	Number of Pages of Zoology	Number of Pages of Physiology and Hygiene	Number of Pages of Biology	Number of Pages of All Text Material	Number of Scientific Terms
Clement ...	288	117	50	95	550	406
Hunter* ...	51	40	137	186	414	340
Hunter† ...	165	130	116	16	427	297
Moon	139	191	139	165	634	518
Peabody and Hunt .	56	155	101	253	565	458
Smallwood, Reveley, and Bailey	180	192	143	150	665	647
Totals	879	825	686	865	3,255	2,666
Average ...	146.50	137.50	114.33	144.16	542.50	444.33

* Hunter's *Civic Biology*. † Hunter's *New Essentials of Biology*.

NOTE: In the above table the scientific terms were counted from the glossary when one was given; otherwise from the index. In counting the number of pages of botany, zoology, etc., the supplementary matter was included in the total.

TABLE 17

DISTRIBUTION OF THE HEALTH CONTENT IN THE TEXTBOOKS OF BIOLOGY USED IN THE COOPERATING SCHOOLS

Author	Total Pages of Health Material	Total Pages of Non-health Material	Total Pages in the Text	Percentage of Health Material
Clement*	121.38	222.62	344.00	35.57
Smallwood, Reveley, and Bailey*	197.54	406.17	603.71	32.72
Peabody and Hunt*	201.65	275.46	477.11	42.23
Moon*	180.59	335.46	516.05	34.99
Hunter (Essentials)*	139.23	272.82	412.05	33.79
Hunter (Civic)	98.08	192.92	291.00	33.70
Totals	938.47	1,705.45	2,643.92	213.00
Average	156.41	284.24	440.65	35.50

* Data from *Health Subject Matter in Natural Sciences* by Claude S. Chappelear. See Bibliography.

CHART III

GRAPHIC REPRESENTATION OF THE AMOUNT OF PAGE SPACE GIVEN TO BOTANY, HYGIENE, ZOOLOGY, AND BIOLOGY IN THE TEXTBOOKS

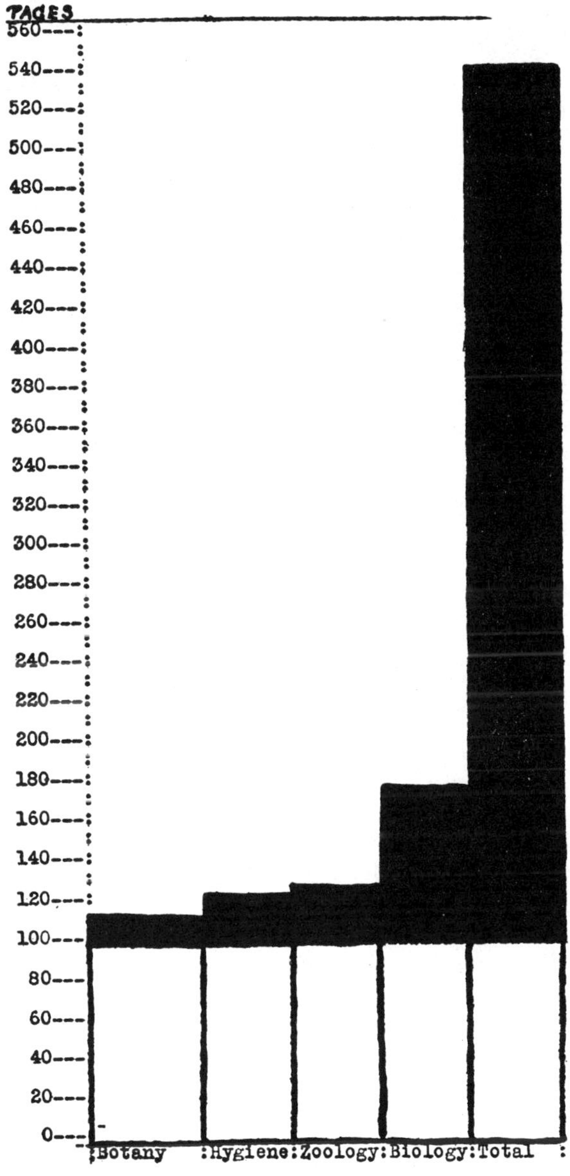

The Analysis of the Biology Laboratory Manuals

The biology laboratory manuals used in the cooperating schools are listed in Table 18. Each manual was analyzed as to the number of units of written work outlined for botany, zoology, human physiology, chemical, and mixed units.

The analysis shows that there are more units of laboratory work provided for botany than for any other division of subject matter of biology. In making this study of laboratory manuals, the manuals of Gruenberg, Atwood, and others were also examined, but in every case the emphasis is on botany, judging by the relative number of exercises, experiments, etc., provided for the study of plants in the laboratory. Human physiology and hygiene in the laboratory manuals rank in importance next

TABLE 18

Analysis of the Laboratory Manuals Used in the Cooperating Schools

Name of Author	Name of Laboratory Manual	Copyright Date	Number of Units of Written Work (Exercises, Experiments, etc.) Classified as to Subject Matter					
			Botany	Zoology	Hygiene and Health	Mixed	Chemical	Total
Blaisdell *	Exercise and Review Book in Biology	1928	24	34	14	4	14	100
Clement	A Laboratory Guide and Notebook in Biology	1925	33	13	9	4	12	70
Moon	Laboratory Manual (Revised Edition)	1927	36	22	10	10	9	87
Hunter	New Laboratory Problems in Civic Biology	1927	80	33	115	30	15	273
	Totals		173	102	148	48	50	530
	Average		43.25	25.50	37	12	12.50	132.5

* The supplementary material in Blaisdell's manual was not included in the analysis.

to botany, judging by the amount of laboratory work outlined in them. There are relatively few exercises and experiments on phases of biology which treat of plants and animals together or in their natural relation to each other.

The laboratory work in biology does not seem to be as well outlined as the laboratory work in physics and chemistry, but the reason for this may be that biology is a much more recent science in the secondary school curriculum. Moreover, biology is usually offered in the ninth or tenth grades, whereas physics and chemistry are generally offered in the eleventh and twelfth grades. Laboratory practices in botany and zoology have been carried over into the field of biology, and this inheritance has greatly influenced the plan and selection of exercises and experiments for the biology laboratory manuals. Many schools do not use a laboratory manual in biology, the teacher outlining the laboratory work required.

The Examination Questions

The College Entrance Examinations

The College Entrance Board of the Middle States and Maryland was organized on November 17, 1900, and the first examination was given in June, 1901. Natural science subjects were not offered in 1901. Botany was the first natural science subject given, being a part of the examination in 1902. Zoology was added in 1907, but the first examination in biology was not offered until 1914.

The pupil has always had a choice of questions in the natural sciences. Beginning in 1914 the questions for each subject were grouped and a still wider range of choice of questions was permitted. The early questions were mostly of the fact or memory question type. The emphasis was on structure in plants and animals until 1914, when the idea of thought-provoking questions began to dominate, and much less emphasis was placed on structure. The emphasis changed from structure to the functional behavior of plants and animals, and at the present time there is a tendency to stress the health and economic phases in the biology examinations. The work of drawings has been emphasized from the beginning, but in the biology examinations offered during the past three years it has been possible for pupils to make a good grade without any drawings at all.

Much credit has been allowed for notebook work in the past. In the first natural science examination—botany in 1902—64 counts were given for examination questions and 36 counts for the laboratory notebook. This proportion for the natural sciences remained about the same until 1911, when a teacher's certificate covering the laboratory instruction was substituted for the submission of a notebook. From 1911 to the present time it appears that the notebook is not counted in arriving at the pupil's College Entrance examination grade, but that the teacher's certificate is required as a part of and a prerequisite to the examination in natural sciences.

Beginning with 1917, health questions appear more frequently in the natural science examinations, and the economic phase receives considerably more attention. Since 1926 there has been a change, in part, to the newer type of questions, such as the true-false, completion, multiple-choice, etc.

In the analysis of the College Entrance examination questions, a study was made of the botany and zoology questions in addition to the biology questions, but biology alone is reported in detail. (See Table 19.) The examination questions were divided into five-year periods, beginning with 1914 for biology, and were classified as fact or memory questions and thought-provoking questions. They were further classified and listed under the headings: plant, animal, chemical and food, human physiology and hygiene, and miscellaneous; also with reference to the requirement of a drawing and the number of questions pertaining to structure.

The Regents Examination Questions

The Regents examination questions were examined in the same manner as the College Entrance examination questions. (See Table 20.) These questions are sent twice a year, January and June, to all the academic schools of the State of New York by the examining department at Albany. The biology questions from June, 1914 to January, 1929 were analyzed.

The letter of inquiry to college registrars regarding the requirements for submitting biology notebooks for admission brought forth such statements as: "We do not at present accept the Regents examination in Biology but we accept the College Entrance Board examination in Biology." Moreover, the notebook

Teacher's Certificate in regard to Laboratory Work in

Biology
(Specify the Science)

This certificate must be filled out in ink and, in every case where a laboratory note-book is required, should be attached to the front cover of the note-book.

Smithville High School
(School)

Smithville, N. Y.
(Post office address of school)

June 15, 1929
(Date)

To Smith College
(University, college, or scientific school that candidate purposes entering)

I certify that during the academic year 19 28—19 29

Sarah Jane Brown
(Pupil's name in full)

while in the laboratory of the Smithville High School
(Name of school)

personally performed and recorded in a suitable note-book 46
(Number of experiments)

experiments not very different from those suggested by the College Entrance Examination Board. This note-book constitutes a true and original record of the experiments.

The time given to the laboratory work has been 39 periods of 45 minutes each, equivalent to 29 periods of 60 minutes each.

The time given to lectures and recitations has been 156 periods of 45 minutes each, equivalent to 117 periods of 60 minutes each.

Half the number of hours given to laboratory work plus the full number of hours given to lectures and recitations is equal to 131½ hours.

(*Signed*) Henry Jones

Teacher of Biology

The teacher may here enter the final grade of 84 per cent.

Laboratory note-books belonging to candidates taking the Board examinations should not be forwarded to the Secretary of the College Entrance Examination Board but if required should be sent directly to the Chairman of the Committee on Admission of the university, college, or scientific school that the candidate purposes entering. The Board is concerned simply in conducting and reporting the results of the actual written examinations.

SPECIMEN CERTIFICATE

was not required, but instead a certificate covering laboratory work when offering biology for admission credit was demanded.

Professor Ben D. Wood states that the function of a college examination is to measure "knowledge of and ability to think in the materials of the course." As proof of such ability the College Entrance Board requires the candidate to furnish a teacher's certificate covering the laboratory work in biology, in addition to taking the examination. The form for this certificate is given below, filled in as though for a single pupil. It is, however, representative of all the pupils submitting notebooks for this study because it represents the average of the data compiled for all.

The College Entrance Examination Board's "Definition of the Requirements" for 1929 specify that the minimum requirements in biology are: "Course I—a year of biology with emphasis on the applications of biology to human welfare" or "Course II—a year of biology with emphasis on the structure and function of plant and animals." The examination questions cover "(1) the structure of plants and animals. (2) the physiology, life history, and classification of plants and animals. (3) the applications of biology to human welfare." The College Entrance Examination Board also specifies in detail just what should be studied in Courses I and II. Course II represents the older type of work and this is the course that most of the pupils who submitted notebooks for this investigation must have pursued.

TABLE 19

ANALYSIS OF THE COLLEGE ENTRANCE EXAMINATION QUESTIONS

Biology: Analyzed as to Subject Matter

Period of Time	Number of Questions on Plants	Number of Questions on Animals	Number of Questions on Physiology and Hygiene	Number of Questions on Chemicals and Food	Number of Miscellaneous Questions	Total Number of Questions
1914–1918	16	12	7	9	20	64
1919–1923	16	19	15	8	14	72
1924–1928	18	18	19	5	15	75
Totals	50	49	41	22	49	211

Biology: Analyzed as to Type of Question

Period	Number of Questions on Structure	Number of Questions Requiring a Drawing	Number of Thought Questions	Number of Fact Questions	Total Number of Questions
1914–1918	6	1	22	42	64
1919–1923	9	11	20	52	72
1924–1928	11	9	39	36	75
Totals	26	21	81	130	211

NOTE: Biology first appeared in the College Entrance Board Examination in 1914.

TABLE 20

ANALYSIS OF THE REGENTS BIOLOGY QUESTIONS

Period of Time	Number of Fact Questions	Number of Thought Questions	Number of Questions on Structure	Number of Questions Requiring Diagrams	Number of Questions on Plants	Number of Questions on Animals
June, 1914, to January 1919	96	54	9	24	51	37
June, 1919, to January 1924	77	73	4	24	54	31
June, 1924, to January, 1929	68	87	7	15	26	32
Totals	241	214	20	63	131	100

(Below is extension of the above table)

Number of Questions on Human Physiology	Number of Mixed Questions	Number of Questions on Health	Number of Questions on Chemicals and Food	Number of Questions on Economic Phase	Total Number of Questions
20	28	50	14	14	150
25	30	62	10	14	150
24	58	64	15	16	155
69	116	176	39	44	455

TABLE 20X

SUMMARY OF HEALTH MATERIAL CONTAINED IN THE NOTEBOOKS, TEXTBOOKS, LABORATORY MANUALS, AND EXAMINATION QUESTIONS

Sources	Percentage of Health Material	Percentage of Non-health Material
The Notebooks (Table 7)	18.37	81.63
The Textbooks (Table 17)	35.50	64.50
The Laboratory Manuals (Table 18)	27.92	72.08
Examination Questions:		
* College Entrance	31.93	68.07
Regents (Table 20)	38.68	61.32
Totals	152.40	347.60
Average	30.48	69.52

* Data from *Health Subject Matter in Natural Sciences* by Claude S. Chappelear. See Bibliography.

Table 20X is a compilation of the health percentages given in Tables 7, 17, 18, and 20, with the percentage of health found by Chappelear in the study of the health content of thirteen College Entrance examination questions. Table 19 gives the result of a study of College Entrance examination questions by years, but does not include specific health data. Table 20X shows that a smaller percentage of health material is found in the notebooks than in the related factors, such as textbooks and examination questions. The probable reason for the relatively small percentage of health material in the notebooks is the fact that laboratory work on plants and animals is stressed more than human physiology and health.

CHAPTER VI

THE QUESTIONNAIRE STUDIES

Questionnaires were sent to the biology teachers, to the pupils who submitted notebooks, and to the principals of the cooperating schools. A letter of inquiry concerning notebook requirements for admission with credit for biology was sent to colleges and universities in New York State and to other educational institutions.

All material in the notebooks was classified and made the basis of Part II of the questionnaire to teachers for their collective evaluation. This part of the teacher questionnaire is shown in Table 28, which gives both a list of the units of written work found in the notebooks and a compilation of the evaluations of the material by the teachers. Along with this questionnaire a personal letter was sent to each teacher. The letter is printed on page 60. Part I of the questionnaire to teachers is given in Appendix A. Tables 21 to 27 give tabulations of the data evolved from the questionnaire to the teachers.

It was necessary to send letters and questionnaires to the principals of the cooperating schools to secure complete data on school facilities for the teaching of elementary biology in the several schools; number of pupils enrolled in the biology classes; list of textbooks and laboratory manuals used in the biology courses; and so on. The information obtained from this questionnaire is tabulated in Tables 29 to 33. The questionnaire is given in Appendix B.

A personal letter and questionnaire were sent to each pupil submitting a notebook for this study in order to obtain his or her reaction to notebook work. Tables 34, 35, and 36 tabulate the information obtained from the pupils. The pupil questionnaire is listed as Appendix C.

For the letter of inquiry sent to college and university registrars and the replies made by them, see pages 90 to 94. The list of colleges and universities to which the letter was sent is printed as Appendix D.

Letter to Teachers

Miss Esther B. Reeve,
Teacher of Biology,
The Central High School,
Smithville, N. Y.

Dear Teacher:

Your school was selected by the State Department of Education last summer as one of fifty representative schools to submit biology notebooks for cooperation in a study of the best practices in biology notebook work. The notebooks were turned over to me for intensive study.

I have analyzed them from many different angles, and from this analysis and that of textbooks, laboratory manuals, examination questions, and other literature, I have developed the inclosed material on which I desire the consensus of opinion of the teachers who contributed notebooks.

Your opinion will help us in evaluating biology notebook work and in formulating suggestions for the improvement of present practices. I shall be very grateful to you indeed for your frank criticism of the inclosed material and high school biology work in general, especially the notebook phase.

Some of the questions and data requested may seem difficult. This is largely due to the fact that I have taken things from the sources as they are—not as you or I probably think they should be. With this in mind, I hope you will answer each question to the best of your ability, writing separately and fully any additional material or comments that you think might be helpful.

Thanking you for your additional cooperation in this study and hoping that our findings, which will be available in the next few weeks, may be of interest and value to you, I am

Very truly yours,

Don O. Baird

The Teacher Questionnaire

The teacher questionnaire consists of two parts. The first part concerns biology notebook work in general. (See Tables 21 to 27, pages 65 to 71.) The second part gives the teacher evaluation of the specific materials found in the notebooks submitted for this study by the fifty cooperating schools. (See Table 28, page 72.) Forty-four teachers took the time to evaluate biology notebook work by answering the questionnaire.

With reference to the pupil's final grade in biology the consensus of opinion is that the recitation should be the basis for most of the grade, the averages for the amount to be given for recitation, laboratory work, notebook work, and objective tests being approximately 2/5, 1/5, 3/20, and 1/4, respectively. The grade on the notebook alone is based, as an average, approximately 2/5 for drawings and 3/5 for the written units of work. Most teachers agreed that very little of the notebook work should be dictated, but several teachers think that some of the drawings may be copied with profit by the pupil.

There is wide variation of opinion among the teachers as to how many teacher demonstrations and pupil demonstrations there should be, and how many of each should be recorded in the pupil's notebook. The same is true of individual experiments. On the whole, teachers think that they should perform two-thirds of all the demonstrations. Practically all teachers indicate that there should be 30 to 35 units of written work in a notebook. Two-thirds of the teachers say that the pupils do not rewrite their notes for the notebook, but most teachers permit the rewriting of the notes if the pupil wishes to do it. There is a wide range in the total number of hours spent by the pupils in writing their notebooks, the least number of hours reported being 27 for the school year and the greatest number 240. The average amount of time, as reported by the teachers, is 74 hours. As an average, the pupil spends in notebook writing 6 hours at home, 10 hours in the school study hall, and 58 hours in the laboratory.

The teachers were asked to rank as to importance five typical units of written work, namely: Oxygen, Osmosis, Photosynthesis, Study of the Frog, and Pulse Rate. Eight teachers ranked Oxygen, 1; fifteen ranked Osmosis, 1; eleven ranked Photosynthesis, 1; three ranked Study of the Frog, 1; and one teacher ranked, Pulse Rate 1. Twenty-eight teachers ranked Pulse Rate as the least important of the five typical units. With reference to the part of biology laboratory work that should be allotted to test for something, as a test for starch; to experience something, as taking one's pulse rate; to draw and write up; carefully observation and write up of natural objects—there is agreement in general as to the fractional part allotted each item. The average fractional part allotted each item is approximately: testing for something, 1/5; experiencing something, 3/10; drawing and write up, 1/5; and careful observation of natural objects and write up, 3/10.

To the question: "In your opinion what percentage of a biology notebook should be based on structure, function, etc.?" the average percentages are approximately: Structure, 18; Function, 18; Economic importance, 12; Adaptations, 20; Life histories, 10; Descriptions, 12; and Comparisons, 10. The notebook objectives, as listed in question 15 of the questionnaire, were not ranked with any degree of sameness, except that twenty-two of the teachers ranked "the development of power of observation" as most important. Nine teachers think "to furnish valuable training in scientific method of thinking" the most important of these objectives.

Many of the teachers consider that biology notebook work contributed to the larger objectives of education, such as health, worthy home membership, etc. Specifically, many teachers said that notebook work does contribute to health through dissemination of knowledge of health; that it makes for cleanliness and sanitation in the home; that it develops responsibility and orderliness; that it creates an interest in birds and outdoor life; that ideals are inculcated; and that such work may lead to an interest in certain vocations as medicine, agriculture, forestry, etc. Even in cases of doubt as to whether notebook work does contribute to the larger aims of education, many teachers say this work should contribute much to the cultural life of the pupil.

More than a third of the teachers reported that field trips are taken to study birds, insects, trees, etc., but only one reported a field trip to gather local material for laboratory study. Twenty-seven teachers require outside reading in science books, magazines, and texts other than the class textbook. Most of these teachers check outside reading by class reports. A few give examinations on outside reading, and some require a written report. Among the supplementary texts used, Moon's *Biology for Beginners* was most popular, but several teachers named texts by Peabody and Hunt, Clement, Hunter, and others. Among the science books other than texts found most useful were: *How to Know Wild Flowers, Life of Pasteur, Birds of New York*, etc. Few magazines were listed, but *Nature Magazine* was mentioned by several.

Two-thirds of the teachers omitted parts of the textbook, the most common reason being "lack of time to study everything." Of the subject matter omitted from the biology course as given in the textbooks, most teachers omit work on ferns, algae, fungi, reptiles, and crustaceans. Several teachers omit the work on eugenics. Two teachers said that they omitted work on eugenics because the pupils were too young for it, and one teacher reported eugenics too advanced. Another teacher said that this topic is beyond freshmen. With few exceptions the teachers use a different order of topics from that given in the textbooks. Many of the teachers use a different order to obtain materials for laboratory work when needed, but several of the teachers had other reasons; for example, to secure a more logical order of presentation of the subject to the pupils. In this connection it is interesting to note that twenty-two of the teachers teach general or synthetic biology, while fourteen teach biology divided into botany, zoology, and human physiology.

Twenty-eight of the teachers say that laboratory exercises should grow out of the class discussions. Twenty-one teachers think there should be no differentiation of class and laboratory work. Thirty-two teachers believe that the type of work should determine whether demonstrations or individual laboratory work is superior. Twenty-two teachers say that the work in biology should take its departure from a course in general science. The teachers are about evenly divided in opinion as to whether there should be drawings of laboratory apparatus. Most of the teachers agree that the drawings should be on a separate page of the notebook and not on the same page as the written matter. Twenty-two teachers are in favor of sex education as a part of the instruction in biology and five teachers think it should be taught as a separate topic. Several teachers say that sex education should not be taught, either as a part of biology or as a separate topic.

Regarding aids to teaching as listed in question 20 of the questionnaire, all but two of the teachers rank fresh materials on plants and animals of first importance and preserved materials second. Charts and microscopes are ranked about the same, but of less importance than pictures, films, slides, and models. Question 22 of the questionnaire—"to list four biologic experiments which you think will develop resourcefulness or initiative," resulted in a great variety of opinions of the teachers. Practically every experiment found in the notebooks was listed as being suitable types for the development of resourcefulness.

Reference to the second part of the questionnaire on the evaluation of specific topics found in the notebooks (see Table 28) will convince the reader that teachers still emphasize structure in teaching biology, both by the fact that the majority of the teachers mark structure studies "A" or "Very Important" and that drawings are required. Of all the topics in the notebooks "Photosynthesis—necessity for light" received the highest number of "A" rank, forty teachers considering this topic "Very Important." Oxygen, under properties of chemicals, ranks next in importance according to the teachers. A number of teachers suggested additional topics which they thought should be in the notebooks, such as "Artificial respiration," "The importance of rest," "Community hygiene," etc.

In connection with the evaluation of units of written work found in notebooks and listed in Table 28, the teachers had been asked to suggest additional units. The following were given:

1. Artificial respiration.
2. Health and sanitation.
3. Comparison of life histories.
4. Importance of living things.
5. Rate of respiration.
6. Disinfection.
7. Importance of rest.
8. Plant culture.
9. Emergency treatment—first aid.
10. Community hygiene.
11. Soils.
12. Difference between physical and chemical changes.
13. Transformation of energy.

Table 28 is of special significance because it gives a list of the units of written work appearing in the biology notebooks four or more times, and because it shows the compiled evaluations of the teachers concerning these units. The table also shows the teacher judgments concerning the desirability of a drawing with each particular unit of written work. There is much more uniformity of opinion on whether or not there should be a drawing than there is on the relative value of the units of written work.

It is also significant that the teachers consider so few of the units of written work "unimportant." Laboratory work receives less than half the stress given the recitation, and the emphasis in laboratory work is on teacher demonstrations. The chief objective of notebook work is to enable pupils to meet college entrance requirements.

TABLE 21

DATA FROM THE TEACHER QUESTIONNAIRE

How much of the pupil's grade should be based on:					*How much of the notebook grade should be based on:*	
Recitation?	Laboratory Work?	Notebook Work?	Objective Tests?	Other Things?	Drawings?	Other Work?
.250	.250	.250	.250	...	.500	.500
.650	.100	.050	.200	...	.333	.666
.333	.166	.166	.333	...	.600	.400
.333	.333	...	.333	...	...	...
.700	.050	.150	.050	.050	.300	.700
.500	.250	.250	...	...	.333	.666
.250	.125	.125	.250	...	.250	.750
.500	...	...	.500	...	.250	.750
.200	.100	.100	.500	.100	.500	.500
.500	.100	.150	.250	...	.400	.600
.250	.250	.250	.250	...	.250	.750
.500	.050	.050	.400	...	.500	.500
.500	.166	.166	...	.333	.333	.666
.500	.125	125	.250	...	.250	.750
...	.400	...	.600	..	.333	.666
.250	.166	.166	.333	.100	.333	.666
.500	.200	.200	.100	...	.333	.666
.600	.200	.200	...	...	.500	.500
.300	.100	.100	.500	...	.500	.500
.500	.250	.250	...	...	.400	.600
.200	.150	.650	...	...	...	...
.400	.125	.125	.350	...	.500	.500
.500	.166	.166	.333	...	.500	.500
.333	.166	.166	.333	...	.333	.666
.500	.200	.200	.100	...	.300	.700
.350	.200	.100	.350	...	...	...
.100	.600	.150	.100	.050	.250	.750
.500	.125	.125	.250	...	.333	.666
.400	.200	.200	.200	...	...	...
.400	.100	.050	.400	.050	.400	.600
.100	.100	.050	.750	...	.500	.500
.250	.125	.125	.375	.125	.600	.400
.600	.200	.100	.150	...	.400	.600
.333	.166	.166	.333	...	1.000	.000
.450	.100	.150	.300	...	.600	.400
.600	.300	.100	...	...	.300	.700
.333	.333	.111	.111	.111	...	...
.500	.250	.166	.083	...	.166	.833
.300	.200	.200	.300	...	.500	.500
Totals 15.265	7.187	5.848	9.867	0.919	13.880	20.111
Average .391	.184	.149	.253	.023	.408	.592

TABLE 22

DATA FROM THE TEACHER QUESTIONNAIRE (*Continued*)

For a year's work in biology state the number of demonstrations and experiments as follows:

Teacher Demonstrations	Pupil Demonstrations	Number to be Recorded	Pupil Experiments	Number to be Recorded	Other Written Units	Number to be Recorded
25	13	38	..	..	..	..
44	2	34	5	27	..	..
..	..	..	15	..	17	10
20	10	30	15	15	..	2
24	12	36	12	12	..	3
19	2	21	4	..	25	25
10	10	15	35	35	12	0
30	15	45	..	..	6	6
20	20	40	20	20	10	..
15	15	30	30	10	25	0
30	0	30	0	..	..	..
34	6	40	6	..	10	10
10	20	30	5	5	25	20
8	17	25	20	20	20	20
30	5	30	10	..	20	10
6	21	27	8	8	..	..
..	..	..	..	..	4	4
..	22	18	..	..	..	..
15	15	30	12	12	..	..
30	20	20	10	10	..	..
5	20	25	15	15	3	3
..	15	15	30	30	0	0
..	3	..	35	35	1	..
18	6	24	8	13	..	..
10	10	20	30	30	..	..
40	..	..	..	..	..	..
40	30	40	20	20	25	0
6	12	18	15	15	..	..
20	10	30	30	30	10	10
Totals 509	331	711	390	362	213	123
Average* 21.20	12.73	28.44	16.95	19.05	13.31	7.69

Average based on actual number reporting each item.

TABLE 23: DATA FROM THE TEACHER QUESTIONNAIRE (*Continued*)

Number of hours spent by pupils during the school year in writing the notebooks:				*Ranking as to importance, from 1 to 5, the following written units of laboratory work:*				
At Home	In the Study Hall	In the Laboratory	The Total Number of Hours	Oxygen	Osmosis	Photosynthesis	The Frog	Pulse Rate
27	0	120	147	1	2	5	3	4
0	0	50	50	1	3	2	5	4
10	130	0	140	1	3	2	4	5
0	0	27	27	1	2	3	4	5
5	5	30	40	2	3	4	1	5
0	..	..	..	3	2	1	4	5
0	24	13	37	1	3	5	2	4
0	0	60	60	1	3	4	2	5
0	0	47	47	4	2	1	3	5
..	..	..	..	2	1	3	4	5
5	0	53	58	3	2	1	4	5
0	0	80	80	2	1	3	4	5
0	24	60	84	4	2	3	1	5
6	0	40	46	3	2	1	4	5
0	0	120	120	3	1	2	4	5
30	30	60	120	5	1	2	3	1
0	0	50	50	4	2	3	1	5
0	0	67	67	2	3	4	5	1
0	0	50	50	3	1	2	4	5
0	0	80	80	1	2	3	5	4
40	0	40	80	2	3	1	4	5
0	5	50	55	3	1	2	4	5
10	10	220	240	3	1	2	4	5
10	10	60	80	4	3	1	2	5
..	..	..	..	1	2	3	4	5
..	..	..	..	3	1	2	5	4
20	20	50	90	3	1	2	4	5
0	5	25	30	3	1	2	4	5
..	..	40	40	2	3	1	4	5
0	0	60	60	3	2	1	4	5
0	35	50	85	3	4	1	22	5
18	18	46	82	2	1	3	4	5
0	0	40	40	3	1	2	4	5
0	0	60	60	5	2	1	3	4
36	36	80	152	3	1	2	5	4
0	0	60	60	2	1	3	5	4
0	0	60	60	4	1	2	3	5
0	0	30	30	4	2	1	3	5
Totals 217	375	1,978	2,547	100	72	86	135	177
Average* 6.38	11.36	58.18	74.91	2.63	1.89	2.26	3.55	4.66

* Based on number reporting each item.

TABLE 24: DATA FROM THE TEACHER QUESTIONNAIRE (*Continued*)

Evaluation of notebook objectives, ranking them 1, 2, 3, etc.:

A	B	C	D	E	F	G	H	I	J	K	L	M	N	O	P	Q	R
3	..	4	..	..	..	..	..	..	7	6	..	..	..	..	3	4	2
3	15	4	6	2	15	1	5	7	14	8	17	12	13	16	9	10	11
1	3	5	7	7	10	2	4	8	13	11	..	2	12	8	6	14	9
1	..	3	2	..	..	..	4	9	5	6	..	..	..	..	7	8	..
2	14	3	4	5	17	1	6	9	10	16	18	11	12	15	7	8	13
3	8	4	6	7	16	5	15	9	14	13	18	10	11	17	1	2	12
.	..	8	10	11	12	5	7	9	18	6	16	15	4	14	1	3	2
1	10	11	9	15	5	2	4	12	13	17	18	14	3	16	8	6	7
1	7	8	5	13	16	4	2	10	11	14	18	6	3	17	12	15	9
1	..	5	..	..	..	8	9	10	11	..	3	4	..	2	6	7	..
1	2	2	1	1	2	2	2	2	3	3	2	2	3	2	2	1	..
1	5	6	7	8	18	2	4	3	12	13	16	14	15	17	9	10	11
3	..	1	..	..	..	..	..	5	9	2	10	..	8	..	4	6	7
1	5	2	8	4	10	3	6	7	11	9	..	6	10	11	7	7	8
1	11	..	4	5	..	9	..	8	10	..	12	15	7	..	2	..	3
2	7	6	5	10	17	3	11	18	8	9	16	8	12	13	1	14	4
1	3	9	5	6	14	..	7	11	12	13	..	17	10	..	2	15	4
1	4	6	2	7	15	5	10	12	13	14	16	7	11	18	8	9	3
1	6	8	..	..	..	..	3	4	..	5	..	..	9	..	..	2	..
9	14	15	..	16	6	12	7	8	4	5	1	6	11	3	10	13	2
1	..	3	2	8	17	7	..	..	4	..	18	16	..	15	5	16	..
1	3	2	10	12	14	11	9	15	5	6	17	..	7	18	4	8	13
1	..	3	5	..	9	7	6	6	..	..	..	5	..	..	2	..	4
2	..	3	..	..	..	1	..	..	..	..	9	5	6	8	..	4	..
3	6	2	15	1	16	9	11	12	13	14	17	..	10	18	7	8	4
4	5	1	6	13	14	7	..	8	10	11	12	1	15	16	3	9	2
1	1	1	1	1	2	1	1	1	2	2	2	18	1	2	1	1	1
1	4	2	6	3	15	5	9	11	10	14	16	2	13	17	8	7	12
2	3	..	..	..	..	3	..	1	..	..	..	..	4	..	1	..	2
3	..	2	4	..	..	..	8	..	5	6	..	4	7	..	1	9	..
2	..	..	..	..	..	..	..	..	..	..	..	3	5	..	1	..	3
2	13	10	9	11	12	8	6	7	16	17	18	16	14	15	1	5	4
1	8	11	6	7	13	9	10	17	14	12	15	9	18	2	5	4	..
2	13	14	12	5	16	1	6	7	8	15	17	9	10	18	3	11	4
1	6	11	14	13	12	4	5	8	15	16	18	3	10	17	2	7	3
1	8	9	2	10	11	12	14	13	11	15	16	..	4	..	7	5	6
1	3	3	..	2	..	3	4	4	8	9	..	3	10	..	4	5	5
1	2	8	12	7	14	5	9	6	15	16	17	11	13	18	4	10	3
	4	7	11	5	8	14	12	10	15	16	17	3	9	13	1	6	2
Totals 67	193	202	196	215	346	171	216	287	349	339	390	257	310	346	165	269	175

Ratios to nearest whole number are:

A	B	C	D	E	F	G	H	I	J	K	L	M	N	O	P	Q	R
2	7	6	7	7	12	5	7	8	10	11	14	8	9	13	5	8	6

A. Development of power of observation. B. To clear up false inferences. C. To fix biologic principles. D. To develop ideas. E. To establish worth of biologic principles. F. To develop skill in biologic drawings. G. To arouse interest in biology. H. To develop power in solving biologic problems. I. To provide techniques for solving science problems. J. As an aid to memory. K. To preserve facts for future reference and review. L. To enable pupils to meet college entrance examinations. M. To develop habits of honesty. N. To develop habits of neatness. O. To offer opportunity for teachers to learn extent of pupil vocabulary. P. To furnish valuable training in scientific method of thinking. Q. To furnish valuable training for subsequent science work. R. To furnish opportunity for the exercise of initiative.

TABLE 25

DATA FROM THE TEACHER QUESTIONNAIRE (*Continued*)

What percentage of the notebook should be based on:

Structure?	Function?	Economic Importance?	Adaptations?	Life Histories?	Descriptions?	Comparisons?
15	15	15	15	15	15	10
10	10	10	10	20	40	0
25	20	5	20	5	10	15
5	5	10	50	10	10	10
10	10	20	15	20	20	5
25	25	5	15	5	15	10
20	20	25	10	10	10	5
15	15	10	20	10	20	10
10	10	10	15	10	20	25
40	10	10	10	5	20	5
15	15	15	15	15	15	10
10	20	20	10	15	15	10
10	25	10	25	10	10	10
10	10	25	10	25	10	10
50	25	0	25	0	0	0
25	15	5	20	10	10	15
5	30	10	30	5	5	15
10	20	10	20	20	10	10
40	35	0	15	0	0	10
20	20	10	20	10	10	10
10	25	0	50	5	5	5
20	20	15	15	5	10	15
30	20	5	20	10	10	5
10	10	10	20	10	10	30
20	40	5	10	5	10	10
20	15	10	20	10	10	15
20	20	10	20	0	10	20
15	15	15	20	15	10	10
10	10	35	30	5	5	5
10	33⅓	8⅓	33⅓	5	5	5
10	10	15	30	25	5	5
30	20	10	10	10	10	10
Totals 575	593⅓	363⅓	648⅓	325	365	330
Average 17.97	18.54	11.35	20.26	10.16	11.41	10.31

TABLE 26

DATA FROM THE TEACHER QUESTIONNAIRE (*Continued*)

Ranking 1, 2, 3, etc., for value in aids in teaching the following:

Charts	Pictures, Slides, and Films	Models	Microscopes	Fresh Materials on Plants and Animals	Preserved Materials on Plants and Animals
5	2	6	4	1	3
3	5	4	6	1	2
5	5	3	4	1	2
6	5	6	4	1	3
5	4	3	6	1	2
5	3	2	4	1	6
6	5	3	4	1	2
4	3	6	5	1	2
2	4	6	3	1	5
4	5	6	2	1	3
5	4	3	6	1	2
3	5	..	4	1	2
4	3	4	5	1	2
4	2	3	6	1	5
4	3	6	5	1	2
3	4	5	6	1	2
6	5	4	3	1	2
4	5	3	6	1	2
3	1	2	4	5	6
5	4	3	6	1	2
4	4	3	2	1	3
4	5	4	3	1	2
3	5	4	6	1	2
5	6	4	3	1	2
4	2	5	2	1	3
1	1	1	3	1	2
2	1	5	6	3	4
4	5	3	6	1	2
2	2	2	4	1	3
3	4	2	5	1	6
6	3	4	5	1	2
6	4	5	3	1	2
6	5	4	5	1	2
5	3	4	5	1	2
3	4	2	3	1	5
6	5	4	3	1	2
4	3	5	6	1	2
Totals 154	139	139	163	43	103
Average 4.16	3.76	3.76	4.41	1.16	2.78

TABLE 27

DATA FROM THE TEACHER'S QUESTIONNAIRE (*Continued*)

What part of biology laboratory work should be:

	A Test for Something, as a Test for Starch?	To Experience Something, as Taking the Pulse Rate?	Drawing, and Writing up the Notebook?	Careful Observation and Write up of Natural Objects?
	.166	.333	.166	.333
	.100	.150	.100	.650
	.300	.250	.300	.150
	.200	.200	.200	.400
	.125	.125	.500	.250
	.100	.700	.100	.100
	.200	.300	.200	.300
	.200	.400	.200	.200
	.200	.300	.200	.300
	.400	.400	.100	.100
	.200	.500	.100	.200
	.125	.500	.125	.250
	.250	.250	.250	.250
	.050	.500	.150	.300
	.200	.300	.250	.250
	.050	.350	.250	.350
	.250	.250	.250	.250
	.200	.300	.250	.250
	.150	.150	.300	.400
	.250	.300	.150	.300
	.100	.500	.100	.300
	.300	.200	.100	.400
	.200	.500	.100	.200
	.333	.333	.333	...
	.200	.400	.150	.250
	.100	.400	.100	.400
	.200	.250	.250	.300
	.100	.100	.500	.300
	.100	.400	.200	.300
	.300	.300	.200	.200
	.150	.500	.150	.200
Totals	5.799	10.441	6.324	8.433
Average	.187	.336	.204	.281

TABLE 28

Data from the Teacher's Questionnaire (*Continued*)

Evaluation of Biology Notebook Work

We have listed below the units of written work appearing in the biology notebooks four times or more.

Please evaluate these units according to the following scale by checking appropriate column.

A—Very important. (May not be omitted)

B—Important. (May be omitted for lack of time; apparatus; or other cause)

C—Unimportant. (Should be omitted)

Under the column headed "drawing" check "yes" if a drawing is desirable for this unit of laboratory work; "no" if not desirable.

Units of Written Work	A	B	C	Drawing	
				Yes	No
I. Chemicals and Food					
A. Composition of air	27	8	4	7	16
B. Composition of water	17	20	2	4	17
C. Properties of chemicals					
1. Oxygen	38	1		16	12
2. Carbon	33	4	2	5	19
3. Nitrogen	31	7	1	10	16
4. Hydrogen	24	11	5	13	14
5. Sulphur	12	20	7	1	21
6. Phosphorus	13	19	8	1	21
7. Carbon dioxide	37	2	1	12	13
8. Oxidation	34	3	1	1	22
9. Chemical elements	17	11	10	3	17
10. Test for minerals	19	14	7	1	19
D. Tests for nutrients					
1. Starch	36	4		2	21
2. Grape sugar	37	3		3	20
3. Protein	38	2		3	20
4. Fats and oils	37	2		1	20
E. Cycles					
1. Oxygen	23	12	1	12	8
2. Nitrogen	23	13	3	12	8
II. Zoology					
A. Protozoa					
1. Paramecium	34	4	1	36	2
2. Ameba	6	24	6	20	5
B. Insects					
1. Grasshopper					
a. Structure	35	3	1	37	
b. Life history	32	7			16
2. Butterfly, life history	24	9	4		15
3. Study of insects	17	10	6	1	11
4. Economic importance of insects	24	3	2	1	15
5. Bee—life history	16	14	8	1	15
6. Fly—life history	22	11	4	2	15

TABLE 28 (*Continued*)

Units of Written Work	A	B	C	Drawing	
				Yes	No
C. Frog					
1. Structure	27	9	3	22	8
2. Life history	30	6	4	8	12
3. Circulation in frog's foot	11	16	11	6	11
D. Fish					
1. Description; feeding habits; respiration	34	5		23	4
2. Life history	29	5	4	2	17
E. Birds					
1. Outline on birds	21	9	4	3	13
2. Description; characteristics	24	11	2	4	14
3. Bird laws	27	10	1	1	15
F. Study of mammals	26	10	3		19
G. Composition of bone	15	14	7	5	13
III. BOTANY					
A. Seeds					
1. Seed dispersal	31	7	2	19	5
2. Types of seeds	25	10	4	15	7
3. The bean seed	33	5	1	34	1
4. The corn seed	30	7	1	33	1
5. Nutrients in bean seed	28	8	4	4	15
6. Nutrients in corn seed	24	11	4	3	15
B. Germination					
1. Necessity of stored food	37	3		7	15
2. Necessity for moisture	35	3		1	19
3. Necessity for air	37	3		6	14
4. Digestion of starch	31	8		2	17
5. Necessity for moderate heat	34	5		1	17
6. Necessity for oxidation	24	7	2	1	13
7. Production of carbon dioxide	30	7	2	5	13
C. Roots					
1. Function	34	4		4	14
2. Structure	29	6	2	24	6
3. Fleshy roots	27	7	5	15	9
4. Root tip	23	11	3	19	6
5. Root hairs	34	3	2	30	2
6. Nutrients	25	9	5	5	13
7. Response to gravity	21	16	3	8	12
8. Response to moisture	17	15	2	8	10
9. Upward course of liquids	32	6	2	22	3
D. Stems					
1. Structure	35	5		21	5
2. Function	33	6	1	10	11
3. Monocotyledonous stems	31	6	3	30	3
4. Dicotyledonous stems	31	6	3	28	3

TABLE 28 (*Continued*)

Units of Written Work	A	B	C	Drawing Yes	Drawing No
5. Response to light	34	4		11	12
6. Rise of liquids	36	4		19	3
E. Study of flowers	34	4		27	4
F. Pollination and fertilization of flowers	32	4	2	10	9
G. The bean plant	17	11	11	13	12
H. The apple	22	9	7	19	7
I. Study of fruits	32	5	2	20	7
J. Leaves					
1. Transpiration	34	5		10	11
2. Function	28	9	1	1	13
3. Structure	31	8		33	
K. Osmosis	37			29	1
L. Photosynthesis					
1. Necessity for light	40			16	6
2. Necessity for chlorophyll	36	3		12	8
M. Bacteria					
1. Study of bacteria	31	7	1	3	15
2. Bacteria; relations to man	34	2	1	4	14
IV. CELLS					
A. Plant and animal	34	5		30	4
B. Study of protoplasm	26	7	3	1	13
C. Likeness of living things	28	6	2	1	15
V. HUMAN PHYSIOLOGY					
A. Food					
1. Digestion	33	3	1	2	14
2. Study of foods	32	4	1		15
3. Digestion of starch by saliva	37	3			18
4. Oxidation in human body	31	4	3		15
B. Organs of excretion	31	5	2		15
C. Narcotics and stimulants					
1. Effect of alcohol	29	7	3		14
2. Effect of tobacco	28	5	4		14
D. Health					
1. Care of eyes	28	4	4		11
E. Respiration					
1. Organs of respiration	30	5	3	3	13
2. Respiration—chest expansion	23	12	3	3	13
3. Carbon dioxide in breathing	26	7	2		16
F. Biology					
1. Definitions	24	9	4		13
2. Noted biologists	27	8	4		14

Letter to Principals

Mr. John Smith, Principal,
The Central High School,
Smithville, N. Y.

Dear Principal:

Herewith find one dollar ($1.00). Will you please have your clerk or a reliable senior student give me one dollar's worth of time in filling out the inclosed form from your files?

Your school was selected last summer by the State Department of Education as one of fifty representative schools to submit notebooks for cooperation in a study of the best practices. The biology notebooks were turned over to me for intensive study.

I have analyzed these notebooks and other literature from many different angles, and desire this additional information to better complete the picture, and help us to better evaluate biology notebook work, and to formulate suggestions for the improvement of present practices in this state.

I shall appreciate any comments you may find time to make regarding high school biology work, especially the notebook phase.

Thanking you for your cooperation, I am

Very truly yours,
Don O. Baird

The Questionnaire to Principals

A résumé of the data from the questionnaire to the principals of the cooperating schools is shown in Tables 29 to 33, pp. 78-85. All the cooperating schools replied to this questionnaire.

The results of this study show that no school offers human physiology in its curriculum. One school offers botany and zoology. Biology is a required subject in thirty-one of the schools. Practically all the schools require two years of science, usually general science or biology and chemistry or physics. A few schools offer physical geography.

Five schools have the biology laboratories separate from the recitation rooms. Four schools have double laboratory periods. The usual number of laboratory periods is 1 to 3 a week. One school has two recitations and three laboratory periods a week the first semester and three recitations and two laboratory periods the second semester. The recitation periods vary in length from 30 to 60 minutes, but the majority of the schools have 45-minute recitation and laboratory periods.

Thirty-eight of the schools have projection lanterns which are used in biology teaching from two to eighteen times a year. Two of the schools do not use their projection lanterns in teaching biology and one school reports that the projection lantern is seldom used. Thirteen of the schools have motion picture machines, but these machines are used very infrequently for instructional purposes in biology. Five schools reported that they do not have an aquarium. All the schools have compound microscopes and, with few exceptions, an adequate number of these instruments for biology laboratory work.

An effort was made to discover how the teachers' class grades compared with the grades awarded in the Regents examinations. No very definite conclusions were possible for biology because so few Regents examination grades were reported; but of a total number of 13,872 pupils taking the Regents examination last year (1928) 6,502, or approximately 40 per cent, failed to pass. These figures are for all biology pupils in the cooperating schools. Teachers who submitted notebooks for this study have a total of 7,545 pupils enrolled in biology. Of this number 1,536, or a little over 20 per cent, failed to pass. Achievement tests are reported by three schools. Twenty schools have records of intelligence tests. Six schools report that they gave the Terman Intelligence test; seven, the Otis test; one, both the Terman and Otis tests; and one each, the Miller, National, Illinois, Multimental, and Dearborn tests. The ratings range from 99 I.Q. to 149 I.Q. The average I.Q. is 113.8.

In forty of the schools the teachers require the laboratory notes to be written during the laboratory period, but the majority of these teachers permit the pupils to rewrite their notes later. There seems to be no standard as to the number of laboratory units of written work required, the range being from 12 to 60 experiments, demonstrations, exercises, etc. The average for all the schools is 35. Twenty-two of the schools reported that field trips are taken in connection with biology laboratory work. Twenty-five schools reported that no field trips are undertaken. The number of field trips taken by schools reporting them ranges from 1 to 6 trips a year, the average number being one trip during the school year. The notebook value measured in terms of the fractional part of the final biology grade ranges from nothing to one-half of the average final grade. The average allowance for the notebook work in all the schools is one-fifth of the final grade awarded the pupil for the biology course.

Practically all the pupils submitting notebooks carried four or more subjects in high school work during the year in which they had biology. The average was four and three-fifths subjects per pupil. The average final grade in all subjects for these pupils is 84.40 per cent. Their average biology grade is 84.45 per cent. The average grade of all the pupils in classes in biology having pupils submitting notebooks is 79.50 per cent. Thus the pupils who submitted notebooks made a slightly better average grade in biology than they did in the average grade of all their studies, and they also made a somewhat higher grade in biology than the average grade of pupils taking biology in the same classes with them.

Twenty-seven of the pupils submitting notebooks took biology in their freshman year; sixteen in their sophomore year; four in their junior year; and five failed to report on this item. The biology enrollment in the cooperating schools ranged from 29 to 884. The percentage of pupils enrolled in biology to the total enrollment ranged from 4.44 per cent to 43.63 per cent. The average percentage of pupils enrolled in biology to the total enrollment for all the schools is 22.20 per cent.

There was a total of 104 biology teachers in the cooperating schools, eighteen of whom are men. Of these teachers, fifty submitted notebooks. The academic qualifications of the fifty teachers submitting notebooks are as follows: B.S., 18; A.B., 20; Ph.B., 1; A.M., 4; M.S., 1; Normal Diploma, 1; and 7 failed to report on this item. The range of experience for these teachers is from 2 to 34 years, the average being 10 3/4 years. The range of years of experience in their present positions is 2 to 29 years, the average being 7 1/4 years. Twenty-nine of these teachers taught biology only. The range of the number of teaching periods for the biology teachers is 5 to 8. The range of biology periods to other subject periods is four and one-half to three-fourths as an average in all the schools.

The foregoing facts show that the biology teachers of New York State are perhaps as well prepared for their work as are teachers in any state. The conditions under which they teach, with reference to laboratory facilities and subjects taught, are no doubt typical of conditions to be found in other states.

TABLE 29

DATA FROM QUESTIONNAIRE TO PRINCIPALS

School marks of pupils in the cooperating schools

Average Class Marks for All Subjects Taken by Pupils Who Submitted Notebooks. (For the year biology was studied)	Average Biology Grade of Pupils Who Submitted Notebooks	Average Grade of the Whole Class in Biology in Each Cooperating School
80	80	71
84	82	83
95	97	84
84	88	84
84	91	78
78	86	79
88	90	84
85	86	72
81	84	75
77	78	83
84	79	85
74	74	78
80	72	78
86	88	88
82	85	77
90	82	78
83	87	75
92	92	..
77	70	81
90	91	91
70	69	80
85	90	81
86	76	..
..	71	71
82	82	77
85	83	72
86	84	80
90	96	86
86	85	86
..	91	..
92	93	77
90	..	..
85	84	79
65	66	..
79	77	74
88	88	73
90	94	79
82	86	75
84	79	75
85	85	..
90	95	96

TABLE 29 (*Continued*)

Average Class Marks for All Subjects Taken by Pupils Who Submitted Notebooks. (For the year biology was studied)	Average Biology Grade of Pupils Who Submitted Notebooks	Average Grade of the Whole Class in Biology in Each Cooperating School
76	86	..
86	84	84
98	97	..
73	79	71
90	87	80
85	85	72
88	93	..
86	88	88
Totals 3,956	4,055	3,180
*Average 84.17	84.48	79.50

* Average based on number reporting each item.

TABLE 30

Data from Questionnaire to Principals (*Continued*)

Data concerning the teachers of the pupils submitting notebooks

Number of Periods Teacher Taught Biology	Number of Periods Teacher Taught Other Subjects	Total Number of Years Experience	Number of Years in Present Position	Total Number of Biology Teachers	
				Men	Women
3	2	14	14	0	1
5	0	8	1	0	2
5	0	5	5	0	1
3	2	24	10	0	1
3	3	34	29	0	1
5	0	3	1	1	1
3	3	3	1	1	0
3	2	2	2	0	1
4	1	11	6	0	1
5	0	16	4	0	2
3	3	3	1	1	0
3	3	3	1	1	0
6	0	10	6	0	2
5	0	..	..	0	1
4	1	3	2	0	1
..	..	22	20	0	1
2	3	4	3	0	2

TABLE 30 (*Continued*)

Number of Periods Teacher Taught Biology	Number of Periods Teacher Taught Other Subjects	Total Number of Years Experience	Number of Years in Present Position	Total Number of Biology Teachers	
				Men	Women
6	0	18	18	1	2
5	0	26	16	0	2
6	0	3½	3	0	1
6	0	3	3	0	1
5	0	..	3	1	1
4	2	7	4	0	1
5	0	9	6	0	2
5	1	18	8	1	1
5	0	14	9	0	2
6	0	4	2	1	1
5	0	24	2	2	7
5	0	4	3	*(2)	(7)
3	2	11	10	0	1
4	1	10	1	0	2
3	..	19	19	1	2
7	0	14	10	0	1
4	2	2	2	0	2
5	0	12	7	0	2
8	0	15	3	1	1
5	0	12	3	0	6
5	0	17	10	0	6
5	0	17	12	0	4
5	0	9	5	1	1
4	2	2½	1	2	0
5	0	..	16½	0	4
5	0	16	8	0	1
5	0	8	8	0	1
4	2	4	4	0	1
5	0	8	8	0	2
6	0	7	4	0	2
4	2	2	2	1	0
5	0	..	25	1	1
5	0	7	3	0	3
Totals 227	37	488	344½	17	82
Average 4.63	0.77	10.60	7.03	0.35	1.67

* Same school reporting two pupils with different biology teachers.

TABLE 31

DATA FROM QUESTIONNAIRE TO PRINCIPALS (*Continued*)

Data on high school enrollment, number of pupils enrolled in biology, number of subjects taken by the students submitting notebooks in the year in which they studied biology, and number of weeks in school year.

Total High School Enrollment	Number of Pupils Enrolled in Biology	Number of Subjects Each Pupil Studied with Biology	Number of Weeks School Was in Session
588	29	4	40
1,151	139	4	38
775	129	4½	38
152	49	6	40
180	56	4	40
850	255	4	40
212	60	5	40
290	72	4	40
650	84	4	40
780	151	5	38
290	60	5	40
290	60	5	40
1,046	350	4	40
1,300	130	4	40
225	78	5	40
209	57	4	38
204	67	3	40
1,139	184	4½	38
497	151	3½	40
763	137	4	40
496	160	4	40
525	176	4½	40
275	120	6	40
*1,007	138	6	40
401	118	5	40
1,096	155	5	38
.....	250	...	40
1,251	164	5	40
2,629	884	4½	40
205	60	5	40
350	100	5½	40
1,202	...	4	37
332	108	5	40
375	80	5	40
1,550	109	5	38
740	200	4	38
2,600	379	5	38
2,008	448	4½	40
619	141	4½	40
329	87	4	38

TABLE 31 (*Continued*)

Total High School Enrollment	Number of Pupils Enrolled in Biology	Number of Subjects Each Pupil Studied with Biology	Number of Weeks School Was in Session
1,249	210	5	38
563	145	5½	40
502	142	4	..
597	66	4	40
203	149	5	*39
688	180	5	40
112	60	5	40
650	151	5	40
1,842	266	3	38
		5½	
		†5	
Totals 35,987	7,545	230	1,892
Average 749.73	157.19	4.69	39.42

* Junior high school.

† Forty-nine schools reporting fifty-one different pupils for this item.

TABLE 32

DATA FROM QUESTIONNAIRE TO PRINCIPALS (*Continued*)

Data on laboratory periods, recitation periods, and amount of credit allowed for notebook work

Length of the Laboratory Period in Minutes	Length of the Recitation Period in Minutes	Number of Laboratory Periods per Week	Number of Recitation Periods per Week	Amount of Credit Allowed for Notebook Work
80	40	2	3	.333
45	45	..	5	.166
..	..	..	5	.050
45	45	..	5	....
40	45	1	5	.200
40	40	1	4	.100
90	45	1	4	.333
45	45	..	5	.333
..	..	..	5	.200
45	45	2	5	.200
45	45	1	5	.200
45	45	1	5	.200
45	45	2	3	....
45	35	..	5	.200

TABLE 32 (*Continued*)

Length of the Laboratory Period in Minutes	Length of the Recitation Period in Minutes	Number of Laboratory Periods per Week	Number of Recitation Periods per Week	Amount of Credit Allowed for Notebook Work
60	60	1½	5	....
50	50	2	3	....
40	40	1	4	....
90	45	1	5	....
45	45	2	3	.250
45	45	2	3	.333
45	45	..	5	....
55	55	..	..	.400
40	40	3	3	.075
50	50	..	5	....
40	40	1	4	.333
45	45	2	3	.333
45	45	2	3	....
45	45	1	5	.250
40	40	2	3	.166
40	40	2	3	.100
40	40	..	..	.250
45	45	3	2	.500
80	40	1	3	.333
45	45	2	5	.333
45	45	2	3	.333
90	45	1	4	....
45	45	2	4	.200
45	45	2	3	.166
45	45	2	3	.100
40	40	..	..	....
50	50	..	..	....
40	40	1	4	.333
..	45	..	5	....
45	45	1	4	.166
45	45	0	5	.333
90	45	2	3	.500
55	55	2	3	.500
55	30	1	5	....
40	40	1	5	....
45	45	2	3	....
..	..	..	5	....
Totals 2,370	2,105	58½	190	8.802
Average 50.43	43.85	1.63	4.04	.259

TABLE 33

DATA FROM QUESTIONNAIRE TO PRINCIPALS (*Continued*)

Data on number of written units of work required for the notebook, number of credits in science required for graduation, and number of pupils passing and failing in the biology course, and in the Regents examination

Number of Written Units of Work Required for the Notebook	Number of Years of Science Required for Graduation	Number of Pupils Failing to Pass in the Biology Course	Number of Pupils Taking the Regents Examination in Biology	Number of Pupils Passing the Regents Examination in Biology
40	1	24	48	27
30	2	9	421	206
30	2	27	234	127
40	2	14	85	46
30	2	3	99	51
30	2	59	370	191
30	..	10	125	60
46	1	9	114	60
45	1	16	140	75
40	1	46	358	194
20	2	3	145	77
20	2	3	442	243
..	2	108	211	105
35	2	47	150	78
40	2	6	108	57
30	1	6	104	55
25	2	28	221	113
36	2	71	213	142
33	2	14	184	99
33	1	20	308	164
35	1	24	210	118
36	2	27	394	213
35	..	48	237	125
42	2	20	291	155
27	2	24	234	117
..	2	20	1,475	765
45	..	6	95	50
35	1	247	154	77
35	1	63	411	268
25	2	5	189	96
30	2	0	148	82
40	2	0	302	170
35	2	2	300	168
35	2	16	1,133	596
35	..	55	1,086	510
40	2	55	460	246
40	..	168	246	129
30	..	47	274	143
45	2	24	464	211

TABLE 33 (*Continued*)

Number of Written Units of Work Required for the Notebook	Number of Years of Science Required for Graduation	Number of Pupils Failing to Pass in the Biology Course	Number of Pupils Taking the Regents Examination in Biology	Number of Pupils Passing the Regents Examination in Biology
33	2	12	171	97
12	1	24	156	88
60	1	34	215	122
50	1	14	278	192
..	2	20	76	46
45	2	5	279	151
30	1	16	514	264
30	1	...		
50	1	37*		
Totals 1,588	69	1,536	13,872	7,369
Average 35.29	1.64	32.68	301.56	160.19

* This column should not be checked against the last two columns because it is not in the same order. The items for the last two columns are data computed.

Letter to Students

Miss Sarah Jane Brown,
Student, Central High School,
Smithville, N. Y.

Dear Student:

Your school was selected by the State Department of Education last summer as one of fifty representative schools for cooperation in a study of biology notebook work.

Your notebook was selected to send to the State Department of Education as representative of the biology work of your school. These notebooks were turned over to me for intensive study. Since you were selected as a representative student of your school, I would like your further cooperation and help on the inclosed form.

Will you please answer each question to the best of your ability, writing on the back any other comments you care to make, and return the papers to me in the inclosed self-addressed envelope, at your earliest convenience.

Thanking you in advance for this additional cooperation, I am

Very truly yours,
Don O. Baird

The Pupil Questionnaire

Forty-seven of the fifty-two pupils submitting notebooks replied to the questionnaire. Of these forty-seven pupils, one had quit school because of ill health and one is now a freshman in normal school. Thirty-six were sophomores the past year (1928-29); five juniors; two seniors; two freshmen.

The pupils say they enjoy laboratory work, the recitation, and notebook writing in the following ratios, respectively: 28–4–14. In other words, 28 pupils preferred the laboratory work; 4 preferred the recitation; and 14 preferred the writing of the notebook. But they thought that they got the most out of laboratory work, the recitation, and writing the notebook in the following ratios: 18–16–12. They were also asked to rank the divisions of biology, as indicated by their notebook work, in order of preference for study, from 1 to 5. Eight pupils gave plants as their first choice, thirteen gave animals; one gave health; eight gave human physiology; and sixteen gave chemical phase. But the ratios of averages of choice were as follows: plants, 2.77; animals, 2.43; health study, 3.88; human physiology, 2.95; and chemical phase, 2.88.

Most of the notebook writing was done at school, and more of it during the laboratory periods than during the recitation periods. According to the pupils most of the written units of laboratory work in the notebooks were teacher demonstrations. Comparatively little laboratory work was done by groups working together on specific problems. A few pupils wrote term papers. Very few class problems or projects were worked out, but two such projects are significant, namely, a study of sewage disposal and the watching of the development of frogs in the laboratory from the eggs to adults. Some of the pupils made special class reports, but mostly on noted biologists, forest conservation, book reports, etc.

Twenty-eight of the forty-seven pupils reporting say that they expect to go to college. Of this number fifteen expect to attend a co-educational institution. Eight pupils state a preference for a women's college and three a preference for a college for men. With reference to the high school subjects that the pupils prefer to study, fifteen state a preference for biology; eight for English; and twelve for mathematics. The subject preferences of the rest are varied. The average number of pupils in the biology classes of which these pupils were members is 27, but the range of size of class is from 6 to 40 pupils.

Several pupils suggested subjects in biology that they thought should have had more time, for example, the study of Mendel's law. One pupil thought that the study of evolution should be omitted from the course in biology. With reference to length of time to prepare lessons in all subjects as compared with biology, fifteen pupils say Latin lessons take more time; seven, algebra; five, geometry; four, English; and one, French. Five pupils state that biology took the most time to prepare.

Practically all the pupils specified particular instances in which knowledge gained from notebook work in biology had proved useful to them. Several found the information useful in the preservation of health and in giving first aid to injured persons. One pupil had saved a drowning person. A number of pupils said they learned how to protect their gardens from destruction by insect pests, and how to select seeds. Some of the pupils found the notebook material helpful in other courses, for example, home economics and physical geography. One pupil found her knowledge of plants and animals gained from notebook writing helpful in her conversations with other people.

Regarding intelligence tests, nine pupils had been informed as to their I.Q. ratings. Of the eleven remaining pupils who had taken intelligence tests and had not been informed as to their I.Q. ratings, the majority made the characteristic reply: "I haven't heard yet." Achievement tests were not reported in sufficient numbers to be significant. (Tables Nos. 34, 35, 36.)

Since more than 92 per cent of the pupils replied to the pupil questionnaire, and many made suggestions relating to subjects they thought should have further study, it is evident that biology is a very stimulating course for high school pupils. Moreover,

TABLE 34

DATA FROM THE PUPIL QUESTIONNAIRE

Work in biology the pupils in the co-operating schools enjoyed most			*Work in biology the pupils in the co-operating school found to be most helpful*		
Laboratory Work	The Recitation	Writing the Notebook	Laboratory Work	The Recitation	Writing the Notebook
28	4	14	18	16	12

TABLE 35

DATA FROM THE PUPIL QUESTIONNAIRE (*Continued*)

Subject matter of biology the pupils liked best as indicated by ranking the items 1, 2, 3, etc.					*Curriculum subject the pupil liked best*	*Curriculum subject the pupil found to take the most time to prepare*
Plants	Animals	Human Physiology	Chemical Phases	Health Study		
1	2	3	5	4	Mathematics	Geometry
3	1	4	5	2	Latin	Geometry
4	5	2	1	3	Mathematics	....
3	1	2	5	4	Biology	Geometry
2	3	1	5	4	Biology	Latin
2	3	5	1	4	English	Algebra
4	1	3	5	2	History	Latin
4	3	2	1	5	Geometry	Latin
3	2	1	5	4	Com. Law	English
3	1	4	2	5	Biology	Algebra
2	3	5	1	4	Biology	Latin
1	5	3	4	2	French	English
2	1	3	5	4	English	Biology
4	3	2	5	1	Art	Algebra
2	3	4	1	5	English	....
4	2	5	1	3	Algebra	Latin
2	1	5	3	4	Manual arts	Algebra
4	2	3	1	5	Biology	Geometry
4	2	3	1	5	Algebra	Algebra
5	3	2	1	4	Algebra	Biology
4	3	1	2	5	History	....
2	1	4	3	5	Algebra	Biology
4	2	1	3	5	English	Latin
4	1	2	3	5	Geometry	English
3	1	5	2	4	Biology	Algebra
1	2	3	5	4	Biology	....
2	1	3	5	4	English	Latin
4	1	2	3	5	Biology	English
3	2	4	1	5	English	Latin
1	2	5	3	4	Algebra	Latin
..	1	..	..	..	History	....
5	4	2	1	3	Biology	Latin
3	4	2	1	5	Biology	Latin
1	2	4	5	3	Biology	Design
1	5	2	3	4	Biology	Latin
4	3	5	1	2	Mathematics	Geometry
4	2	1	3	5	Chemistry	French

TABLE 35 (*Continued*)

Subject matter of biology the pupils liked best as indicated by ranking the items 1, 2, 3, etc.					*Curriculum subject the pupil liked best*	*Curriculum subject the pupil found to take the most time to prepare*
Plants	Animals	Human Physiology	Chemical Phases	Health Study		
2	1	4	3	5	Biology	Biology
2	5	4	1	3	Geography	Latin
1	3	2	4	5	Biology	Latin
4	3	1	5	2	....	Biology
3	2	1	4	5	English	....
2	4	3	1	5	English	....
3	4	1	5	2	Biology	....
2	3	4	1	5	Mathematics	Algebra
1	3	5	4	2	Algebra	Latin
Totals 125	112	133	130	176		
Average 2.78	2.43	2.96	2.89	3.91		

some of these pupils put to practical, everyday use knowledge gained from the study of biology, as is clearly shown by their statements concerning first aid, care of gardens, etc.

Even though the teachers gave more than twice as much emphasis on the recitation as on the laboratory work, Table 34 shows that the pupils prefer laboratory work. Moreover, the majority of the pupils believe that the laboratory work was the most helpful to them. With reference to the subject matter of biology which the pupils liked best, the author thinks it is significant that health study was ranked the lowest. Evidently the health phase of biology was not presented to the pupils in the most interesting manner or, as is probable, adequate laboratory facilities for the study of health topics were not available in the several schools. It is also significant that the majority of the pupils like the study of plants the best. This may indicate that biology teachers are especially well prepared to teach the plant phase of biological subject matter.

TABLE 36: DATA FROM THE PUPIL QUESTIONNAIRE (*Continued*)

Data concerning the number of pupils in the biology classes of the pupils submitting notebooks in each of the cooperating high schools.

First Semester		Second Semester	
(1)	(3)	(2)	(4)
25	26	23	23
25	20	25	20
34	27	33	31
36	35	35	40
30	29	30	29
16	20	14	17
32	30	32	25
30	30	30	35
30	7	30	6
32	18	32	24
34	30	31	30
38	25	40	35
28	22	30	20
26	20	21	30
25	35	26	27
38	30	40	30
20	21	20	17
26	25	24	26
30	27	30	29
28	23	29	25
30	25	34	25
Totals	1,138	Totals	1,153
Average	27.09	Average	27.45

The Letter of Inquiry to Registrars

The following letter of inquiry was sent to registrars of colleges and universities in New York State, and to other educational institutions, concerning notebook requirements when offering biology for admission credit:

Letter to Registrars

The Registrar,
Ohio State University,
Columbus, Ohio.

Dear Sir:

Please send me a statement of your entrance requirements and a catalog of the college of Science and Arts.

I wish to present a year's work in biology for admission credit. Is a notebook required? If so, should I bring my notebook with me or mail it to the university examiner?

Very truly yours,

Name

To this letter of inquiry the response was prompt and in most replies specific information was given, but in two cases the registrars advised that no information could be given regarding admission until the applicant submitted a transcript of his secondary school record. Excerpts from the replies of the registrars giving specific information are:

I regret that Biology cannot be accepted as an entrance credit. Either Physics or Chemistry may be offered as the required science.

We do not at present accept the Regents examination in Biology but we accept the College Entrance Board examination in Biology. The Biology course is approved only after a satisfactory notebook is submitted to us.

If the course in Biology which you wish to offer for admission is the course given under the Regents Board of the State of New York I would say that this course does not meet our requirements for an entrance unit. If, however, the course which you have taken is similar to the one outlined by the College Board this will be quite satisfactory. It will not be necessary for you to present a notebook in this subject but you should have the inclosed form filled out by your instructor and returned to us.

Biology is one of the subjects accepted for admission to ——. We do not, however, require a notebook to be presented. All that we ask is that your instructor in Biology shall send us a statement that you have performed the necessary experiments.

I would advise that the submission of a notebook in any of the sciences is no longer required, although the preparation of such a book is, I presume, required at all first-class schools.

Biology is one of our regular entrance subjects. It is not necessary to present a notebook in Biology.

In regard to your Biology notebook, if you are studying in an accredited secondary school we shall accept the school's statement of your work. We shall, however, be glad to see the notebook which could be sent in at the time when you fill in your application papers.

It would be necessary for you to bring your notebook to get full credit for your Biology.

We do not require the candidate to submit the science notebook even though science is offered for entrance.

If your Biology is affiliated and your State University will accept it for entrance, it will be accepted here without any notebook whatsoever. All that will be necessary will be for your superintendent to send a complete official transcript of your high school work to the Registrar of —— College.

If you plan to present for admission to —— College your year's work in Biology by passing a —— University examination in that subject, it will be necessary for you to hand in as a part of the examination a notebook certified by your teacher.

If you are attending a school which is on the list of schools approved by —— University for the certificate privilege and your principal is willing, on the basis of your grade in Biology, to certify you in that subject, or if you pass a Regents Examination in Biology with a grade of seventy-five or higher, it will not be necessary for you to present your notebook to the college.

It will be necessary for you to present an official record of your entrance credits before any decision can be rendered about your admission.

If you plan to enter —— College by certificate and present Biology as one unit, it will not be necessary for us to see your notebook. If you enter by College Entrance Board examinations or Regents examinations, the same thing is true.

I can give you no definite information concerning your Biology until we have official transcripts of your record. The notebook will probably not be required.

If you have taken the preparatory work in an approved school it will not be necessary for you to submit your Biology notebook.

A student is not required to present the notebook of her laboratory work in Biology. She should, however, present us with a statement from her teacher that she has completed the requisite number of experiments.

As you will see by reference to the catalogue, we do not accept the Regents course in elementary Biology unless the College Board examination is passed in that subject. In this case, it is necessary to submit the notebook for the approval of the departments here.

Regarding Biology, would say that this subject is not required for entrance here except as it is required or elected for high school graduation. It is accepted for entrance as one unit of the 15 required.

You will find however that the Regents examination in Biology is not accepted for entrance and Biology counts as an entrance unit only when a college board examination is taken in this subject.

It will not be necessary for you to submit your Biology notebook as I presume that credit for the same has been given you in the final mark in that subject.

If you have graduated from an approved school and present New York Regents or other satisfactory credit in Biology, it will not be necessary for you to submit your notebook.

If you wish to bring your notebook in Biology we shall be glad to review it.

If your work in Biology is recommended, it will be satisfactory if you bring your notebook when you come.

It will not be necessary to present your notebook. Your school record will be sufficient.

We do not require Freshmen to present a notebook in Biology, but only the official statement from the Principal of the High School to the effect that the course prescribed by the Syllabus of New York State has been completed.

It is not usually required that a student offering Biology credit submit a notebook in that subject to this office. However, we can state this definitely after a record of your work has been received at this office.

It will not be necessary for you to submit a notebook in Biology if your work is accredited by the high school.

A Biology notebook is not required for presentation when offering a unit in Biology. A standard course from any approved school where laboratory work is given is accepted provided a satisfactory grade is secured.

Applicants offering Biology are not required to take a laboratory examination or to submit a laboratory notebook. They must, however, submit a laboratory certificate such as I enclose.

Relative to your question concerning credit for Biology, permit me to state that if you have satisfactorily completed said course in an accredited secondary school you will be given full credit for the same.

We will accept a year's work in Biology for the one unit required in science. If you receive a certificate from your principal in this work it will not be necessary for you to present your notebook.

It is not necessary that you bring your Biology notebook with you if the school is willing to include this subject in its certificate.

If you have taken your preparatory work in an approved school, it will not be necessary for you to submit your Biology notebook.

If you are a graduate of an accredited high school in the State of New York and have passed the Regents examination in fifteen units of high school work in acceptable subjects, we shall be glad to consider your application for admission to the . . . State University . . . I suggest that you bring your notebook with you."

—From a mid-western state university.

In keeping with the best educational thinking of the day, and as a result of careful experimentation over a period of four years, the Colorado State Teachers College has decided to admit any high school graduate who is recommended by you (High School Principals and Superintendents) who has health, character, and ability to do college work.—George Willard Frasier, President, Colorado State Teachers College.

In this list of replies from registrars giving specific information regarding notebook requirement for admission credit, one college will not accept biology at all; one is willing to examine a notebook if submitted; one is willing to review a notebook; and only two require the submission of a notebook. In many colleges and universities a certificate is assumed to cover laboratory notebook work in biology, and a few colleges are not concerned with notebooks at all.

Bennington College, Bennington, Vermont, has not yet opened its doors to students, but the letter of inquiry concerning admission was not disregarded. The present emphasis on biology in college is strikingly brought to light by this new women's college. The sponsors welcome

The opportunity to start afresh on clear ground, guided solely by the experiences of the last half century, by the needs of the students to be educated and by the demands of the modern world on women is great enough to stir the imagination of even the most cautious.

The tentative curriculum of Bennington College gives at least one-third of the time to the study of natural science, including biology.

The Sarah Lawrence College, Bronxville, New York, was founded two years ago as a junior college for women. The trustees of this institution state that "its aim will be to give two years of college work of the quality equal to that of the best colleges for women so far as concerns the standard of scholarship, the method of instruction, and the satisfaction by examination of the completion of courses." It is interesting to note that the authorities of this college also state that "in natural science, the course will be descriptive and informational rather than experimental. An adequate background for more advanced work will be given, and an understanding of the chief generalizations will be sought."

More than half of the pupils submitting notebooks expect to attend college upon graduation from the secondary school, and in view of this fact and the fact that they will offer biology for admission credit, notebook work as evaluated by the admission authorities of colleges and universities is of considerable importance. But the main purpose in obtaining statements from college registrars concerning the submission of a laboratory notebook when offering biology for admission credit was to ascertain the present practices in this regard and to discover what degree of uniformity of requirement existed.

It is evident from the replies received that in most instances the submission of a notebook is not a requirement for admission credit, but that it is assumed that all secondary schools offering biology require laboratory notebook work, and that the certificate of the school covers this phase of the work.

Courses of Study

Twenty-seven of the cooperating high schools submitted copies of their courses of study. Several of the schools state that they follow the Regents syllabus. A few schools report that they are now reorganizing their courses of study.

None of the courses of study submitted give any data on notebook work in biology, but in several of the courses of study the educational value of a course in elementary biology is made a matter of comment. Every school requires biology in the first year of the college preparatory courses.

The specific requirements regarding a course in biology in the schools submitting courses of study are given below; but in some cases it was not quite clear whether biology is absolutely required or just expected as a customary thing.

Table 37 gives the typical general requirements in the ninth grade for all courses of study in the high schools.

TABLE 37

TYPICAL COURSES IN THE 9TH GRADE OF CERTAIN HIGH SCHOOLS

Business Course	Secretarial Course	Science Course	College Preparatory Courses
English Commercial Arithmetic Biology or General Science Elementary Business Training	English Biology or General Science Elementary Business Training or Civics Commercial Arithmetic	English History Biology or General Science Elementary Design or Elementary Repair Elementary Business Training or Elementary Design	English Biology Latin or History Civics Drawing

CHAPTER VII

SUMMARY AND IMPLICATIONS

Summary

The results of the investigation of biological laboratory notebook work in New York State may be summarized as follows: the notebook work has followed the traditional lines of the older sciences—chemistry, physics, botany, zoology, and human physiology. Much of it appears to be too formal and does not seem to have kept pace with the rapidly changing needs of society. The notebooks show that practically no effort is made to use either them or the laboratory as training for the use of leisure time—the collection of plants, specimens of local fauna, insects, etc., or the taking of field trips for pleasure. The notebook work does not seem to offer any prevocational training interests. A feeble effort is made to teach health through laboratory experiments and demonstrations, as is indicated in the notebook records, but the major emphasis is on plant and animal structure. The functional nature of biology is not greatly stressed—at least not in notebook work. There seems to be no uniformity in the use of drawings in the notebooks.

Implications

I

John Dewey[1] in *School and Society* stated that

> The chief source of waste in American education is failure at all stages from the elementary school through the college to utilize the experiences of children and youth gained out of school. Isolation of school and life is the chief cause of harmful isolations within the school. The problem could be best approached as one of continuous differentiation rather than as one of external articulation.

One way to utilize the experiences which high school boys and girls gain outside the schoolroom is through notebook work in

[1] John Dewey, *School and Society*, 1899, p. 78.

biology. The notebooks could become records of field trips, health practices, profitable uses of leisure, and many other out-of-school activities in the field of biology. The notebooks were examined with these possibilities in mind.

II

The pulse rate experiments in the notebooks examined are not definitely related to blood pressure and body temperature nor associated with the degree of health one enjoys. They might be more meaningful if they were. Some pupils, perhaps, would know that the temperature of the body and the circulation of the blood vary greatly among different animals. Many pupils would know, for example, that reptiles are cold blooded and their circulation is slow, while birds have rapid circulation and high body temperature. In this connection pupils might be encouraged to discover why the artificial incubation of hen's eggs requires a temperature of 103 degrees Fahrenheit.

III

President Nicholas Murray Butler states that "it will be more important, one of these days, to educate men and women to use their leisure time than it will be to educate them for an occupation." The notebooks have very little material in them bearing directly on the profitable use of leisure time, but there are many exercises which would be more or less suggestive in this regard. The study of the fish, if properly directed, might well lead the pupil to an interest in game fish and fishing for health and recreation. Every notebook contains some material on birds. Bird study, if properly directed, might lead pupils to spend their leisure time in the parks and fields with the bird glass and camera. In a few notebooks, references to sportsmen are found, but no distinction is made between sportsmen as hunters and miscreants who kill song birds. There should be a very clear distinction in the minds of the pupils between hunting game birds and animals and the thoughtless destruction of wild life.

Every notebook contains drawings and it would seem that an interest in drawing might suggest sketches of streams, trees, animals, etc., as a fine art and a worthy use of leisure. Biology teachers should also stress ideals to pupils, showing, for instance, that nature is background of all art.

IV

Doctor Otis W. Caldwell, writing in *School Science and Mathematics*, February, 1921, stated that:

"Science opens the door to many useful and pleasurable avocations. For example, photography may be taken up by many but most intelligently by one who understands something of the nature of light, the action of lenses, the chemical changes involved in exposing, developing, and fixing plate or print. In the city and in the country, at the seashore, mountains, and elsewhere, nature is prodigal of her store of wonders. If the natural interest in these things has been developed and deepened by elementary courses in biology, botany, or zoology, not only is there added pleasure and enjoyment, but the door has been opened to wider interests and to a rapidly growing fund of valuable literature regarding science."

The word for "leisure" in Greek is the same as for "school," since "schola" meant leisure, but we moderns got away from the idea of school as a place to prepare ourselves to enjoy life more fully. We became so involved in curriculum of occupation that we lost sight of the fact that children have a life to live.

Many of the notebooks refer to field trips, visits to a museum of natural history and local places of interest, but nothing is said of hiking for fun and for health. Camping, care of pets, horseback riding, photography, swimming, etc., are all out of the picture so far as the biology notebooks are concerned. Sports for thrills play an important part in our national life, and have a biologic basis for their enjoyment; yet no notebooks have even a reference to sports of this type. The care and feeding of pets would seem to be a fruitful field for purposeful activity in the laboratory but seldom is the subject mentioned.

V

Next to the worthy use of leisure, prevocational training should offer the greatest incentive to pupil interest in biology and notebook work. Even young children realize that some day they must engage in productive activity—that their parents will not always support them. Every normal child has a natural interest in earning money, so universal is this appeal. Thus this natural interest may be made a real life situation to the pupil by providing an occupational incentive for the study of biology.

Biology, in the hands of a skillful teacher, is probably as suggestive of suitable life careers as any subject taught in the secondary school curriculum. A master teacher may easily, in his teaching, point out to boys and girls the many desirable occupations directly and indirectly concerned with biologic science. Some occupations may be made the basis for laboratory investigations on a high school level of appreciation and understanding. For example, public health work in the school; taxidermy; chicken hatchery project; etc. In many cases pupils would feel that they are engaging in purposeful activity if such laboratory projects were carried out as prevocational training as well as routine work in biology.

VI

No one would recommend that high school boys and girls engage in microscopic study of active germs of human diseases, but studies could safely be made of the cause and prevention of common colds which are responsible for most absences from school. Feeding experiments could be carried out on small animals; health surveys of the school and community could be made; the purity of the local milk and water supply could be investigated; health records of individual pupils of the biology class could be kept as laboratory exercises; etc. Such laboratory work would be purposeful activity on the part of the pupil and would result in more effective notebook work.

VII

The evaluations which the teachers made of the notebook work indicates that they recognize its importance and do attempt to get good work done; but in many instances they teach other than biology subjects and in some cases do not have sufficient time for effective laboratory teaching. Yet some of the teachers say that they do not use any particular textbook and that they write their own laboratory outlines. All of them have had sufficient training to be well qualified teachers of biology, but technical knowledge alone is not enough. There must be a keen interest in the broader problem of biology teaching—an interest in the subject for its functional value in human society.

VIII

The study of the related factors in biology notebook work, such as the influence of textbooks, laboratory manuals, examination questions, syllabi and courses of study, raised some serious questions.

On the whole, the authors of the textbooks and laboratory manuals used in the cooperating schools agree that biology should be taught in its functional relationship to man, and that structure should not be emphasized; yet the structure of plants and animals is stressed in the notebooks. One reason for the emphasis on formal requirements is the Regents and College Entrance examinations, and the requirement of the state syllabus for biology. These however serve as clubs—not as bases for points of departure. Likewise, the courses of study are written with the dominant idea of pupil success in formal examinations.

Comparatively few of the notebooks use the "report" form of writing records of laboratory work. This form is best in every case where such material can be written in a clear and simple manner since it permits a more natural division of the subject and also promotes the use of good English and originality. It avoids, moreover, set expressions and stereotyped forms. This is a valid reason for discarding the printed type of notebook which require merely the filling in of blank spaces.

The analysis of examination questions shows that there has been considerable improvement in type and scope in the past few years, but these questions do not pretend to measure attitudes and ideals, or to promote the study of the things of life considered to be most worthwhile for the great majority of high school pupils.

The examinations tend to become the motivating force in the study of biology, and cause the real purpose and aim of the teacher to be to prepare the pupils for the examination rather than to prepare them to meet successfully the great social problems of life.

It is certainly unscientific to conduct experiments and to teach biology to satisfy just one phase of the work, namely, the passing of a set examination. It is like trying to isolate one phase of a human being's life from the whole of his life. Very few colleges require the submission of a notebook for admission credit. Practically all of them accept a certificate covering laboratory work, and that being the case, it seems quite possible to make the laboratory work much more related to the life of the pupil.

IX

In submitting to the teachers all items that appeared four times or more in the notebooks, the question was asked: "What additional units would you like to add?" Very few teachers named additional units. This would seem to indicate that teachers are either satisfied with the present subject matter units, since very few were considered as unimportant, or that they have no clear plan as to what changes they think advisable. But if their pupils must continue to take the Regents examinations and be successful in passing them, they must follow the syllabus slavishly. Under the present system forty per cent of the biology pupils in the schools cooperating in this study who took the Regents examinations last year failed. This would indicate that there is need for a change for the better in the present system of examinations. Still less do they measure the degree to which what is learned enters the life of the pupil to remake it to a higher and richer level.

X

It is the belief of the author that a few major projects in biology laboratory work planned by the teacher for a year's course would result in activities more valuable to a pupil than thirty or more isolated experiments and demonstrations. For example, the pupils might carry out a health project for the entire school year, keeping accurate records of the changing environment and their personal habits in relation to their health. Each pupil could select with the assistance of the teacher the things he wishes most to do and to be responsible for during the school year. Pets could be selected and cared for during the school year, an accurate record being kept of their feeding habits, special characteristics, health, etc. Various plants could be grown in the laboratory, home and school gardens, and experiments made with them over a period of several months. All outdoors could be made a laboratory for boys and girls by training them to enjoy hikes, skating, fishing, etc., but always with intelligent observation of the marvels of nature about them.

In brief, the author would have the work of high school biology motivate the lives and activities of the pupils in purposeful activity rather than be a course in a prescribed task of writing so many units of work in a notebook.

BIBLIOGRAPHY

ATWOOD, W. H. *Biology.* pp. XI—522. P. Blakiston's Sons & Co., Philadelphia, 1927.

ATWOOD, W. H. *Projects, Problems, and Experiments in Biology.* pp. VI—106. Blakiston's Sons & Co., Philadelphia, 1927.

BAILEY, G. A. and GREENE, R. A. *Laboratory Manual.* pp. ix-142. Allyn & Bacon, New York, 1922.

BLAISDELL, J. G. *Exercise and Review Book in Biology.* pp. VIII—152. World Book Co., Yonkers-on-Hudson, N. Y., 1928.

CALKINS, N. A. (Chairman). *Committee of Ten. Report on Secondary Schools.* pp. XII—249. American Book Co., New York, 1894.

CHAPPELEAR, CLAUDE S. *Health Subject Matter in Natural Sciences.* Bureau of Publications, Teachers College, Columbia University, New York, 1927.

CLEMENT, A. G. *Living Things.* pp. VIII—488. The Iroquois Publishing Co., Syracuse, N. Y., 1924.

CLEMENT, A. G. *Suggestions for Teaching Biology.* pp. 44. The Iroquois Publishing Co., Syracuse, N. Y., 1924.

Cardinal Principles of Secondary Education. pp. 32. Bulletin No. 35. U. S. Bureau of Education, 1918.

CLEMENT, A. G. *A Laboratory Guide and Note Book in Biology.* pp. XII—90. The Iroquois Publishing Co., Syracuse, N. Y., 1925.

DOWNING, ELLIOT R. *Teaching Science in the Schools.* pp. XIII—185. The University of Chicago Press, Chicago, Ill., 1925.

GEER, WILLIAM C. "The Teaching of Chemistry in the Secondary Schools." *School Review,* December 28, 1906.

GRUENBERG, B. C. *Biology and Human Life.* pp. XIV—592. Ginn & Co., New York, 1925.

GRUENBERG, B. C. *Manual of Suggestions for Teachers.* pp. IV—95. Ginn & Co., New York, 1919.

HUNTER, G. W. *New Essentials of Biology.* pp. X—453. American Book Co., New York, 1923.

HUNTER, G. W. *New Laboratory Problems in Civic Biology.* pp. 283. American Book Co., New York, 1916.

HUNTER, G. W. *New Civic Biology.* pp. IX—448. American Book Co., New York, 1926.

KINSEY, A. C. *An Introduction to Biology.* pp. XI—558. J. B. Lippincott Co., Philadelphia, 1926.

KINSEY, A. C. *Field and Laboratory Manual in Biology.* pp. XIX—151. J. B. Lippincott Co., Philadelphia, 1927.

LINVILLE, H. R. *The Biology of Man and Other Organisms.* Harcourt, Brace & Co., New York, 1923.

Locy, W. A. *Biology and Its Makers.* pp. 469. Henry Holt & Co., New York, 1908.

Moon, T. J. *Biology for Beginners.* pp. viii—647. Henry Holt & Co., New York, 1926.

Moon, T. J. *Laboratory Manual.* pp. 231. Henry Holt & Co., New York, 1927.

Peabody, J. E. and Hunt, A. E. *Biology and Human Welfare.* pp. xii—585. Macmillan Co., New York, 1924.

Powers, S. R. "Vocabulary of Scientific Terms for High School Students," *Teachers College Record,* November, 1926.

Reorganization of Science in Secondary Schools. pp. 62. Bulletin No. 26. *U. S. Bureau of Education,* 1920.

Syllabus for Secondary Schools (Biology). pp. 25. Department of Education, Albany, N. Y., 1925.

Smallwood, W. M., Reveley, I. L., and Bailey, G. A. *New Biology.* pp. xix—704. Allyn & Bacon, Chicago, 1924.

Smallwood, W. M. *A Text Book of Biology.* pp. 393. Lea & Febiger, Philadelphia, 1924.

Scott, George G. *The Science of Biology.* pp. xii—617. Thomas Y. Crowell Co., New York, 1925.

Trafton, G. H. *Biology of Home and Community.* pp. x—614. Macmillan Co., New York, 1923.

Thorndike, Edward L. *The Teacher's Word Book.* pp. 134. Bureau of Publications, Teachers College, Columbia University, New York, 1921.

Woodruff, L. L. *Foundations of Biology.* pp. xxiii—546. Macmillan Co., New York, 1927.

APPENDIX

APPENDIX A

(Questionnaire to Teachers—Form A, Part I)

BIOLOGY NOTEBOOK WORK

Please answer the following questions concerning practices in biology notebook work in the schools selected for this study. If additional space is required use the back of the sheet.

1. How much of a pupil's grade, do you think, should be based on th following:
   Recitation?________ Laboratory work?________ Notebook? ________
   Objective tests?________ What other things, if any?________
2. How much of a pupil's notebook grade should be based on drawing?________ On written work?________
3. State approximately for a year's course:
   a. The number of demonstrations you think the teacher should perform ________
   b. The number of demonstrations you think the pupil should perform ________
   c. The number of demonstrations you think should be recorded in the pupil's notebook________
   d. The number of experiments you think the pupil should perform________
   e. The number of experiments you think should be recorded in the pupil's notebook________
   f. The number of other units of written work; as: exercises, studies, problems and projects, you think the pupil should perform________
   g. The number of other units of written work, referred to in *f.*, you think should be recorded in the pupil's notebook________
4. What part of the notebook should be dictated?________ What part copied from the text or other sources?________
5. List two things, if any, that should be dictated: (1)________ (2)________ Two that should be copied: (1)________ (2)________
6. Estimate the time spent, in hours, by the pupil on notebook work: (1) At home________ (2) In study hall________ (3) In laboratory________
7. Do your pupils rewrite their notes for their notebooks?________
8. How many field trips did your class make last year?________ What was the purpose of two or three of your field trips? (1)________ (2)________ (3)________

9. List the supplementary biology texts you find most helpful in your teaching: (1)__________ (2)__________ (3)__________ (4)__________
10. List the supplementary biology manuals you find most useful in your laboratory teaching: (1)__________ (2)__________ (3)__________ (4)__________
11. Did you cover the whole text book last year?__________ If not, what parts of the text did you leave out?__________ Why?__________
12. Did you find it advisable to change the order of the topics in the text book?__________ What order did you follow?__________ Why?__________ Do you use different orders in different years?__________
13. Do you teach general or synthetic biology or is your course divided into botany, zoology, and human physiology? Underscore either "general" or "divided" as the case may be.
14. Rank as to importance the following units of laboratory work, as 1, 2, 3, 4, and 5:

	Rank
a. Oxygen	
b. Osmosis	
c. Photosynthesis	
d. The frog	
e. Pulse rate	

15. In your opinion what percentage of a biology notebook should be based on:

	Per Cent
a. Structure	
b. Function	
c. Economic importance	
d. Adaptations	
e. Life histories	
f. Descriptions	
g. Comparisons	

16. Please evaluate these biological notebook objectives, marking the best 1; second best 2; and so on. (If you wish, add other objectives you believe desirable.)

Objective	Rank
a. Development of power of observation	
b. To clear up false inferences	
c. To fix biologic principles	
d. To develop ideas	
e. To establish worth of biologic principles	
f. To develop skill in biologic drawings	
g. To arouse interest in biology	
h. To develop power in solving biologic problems	
i. To provide techniques for solving science problems	
j. As an aid to memory	
k. To preserve facts for future reference and review	
l. To enable pupils to meet college entrance examinations	
m. To develop habits of honesty	
n. To develop habits of neatness	
o. To offer opportunity for teachers to learn extent of pupil vocabulary	
p. To furnish training in scientific method of thinking	
q. To furnish valuable training for subsequent science work	
r. To furnish opportunity for the exercise of initiative	
s.	
t.	

17. Write "Yes" or "No" before each of the following statements to indicate your agreement or disagreement.

________(1) Laboratory exercises should grow out of problems raised in class discussions.

________(2) It is desirable to differentiate clearly between laboratory work and class work.

________(3) The type of work should be a basis for determining whether the demonstration or individual laboratory experiment method is superior.

________(4) The work in general biology should take its departure from general science.

________(5) The requirement of drawings of laboratory apparatus is advisable.

________(6) Drawing should be on separate pages from the written work.

(NOTE: If you wish to qualify your answer to any of these statements, do so on the back of this sheet, numbering your statements to correspond with the above statements.)

18. Do you require outside reading?________ In other texts?________ In other science books?________ In magazines?________ How do you check the pupil on this: Examination?________ Class report? ________ Paper?________ ________ ________

19. List four science books (other than texts) you find most useful in your teaching.____________________
    List four magazines you find most useful.____________________

20. Please rank the following in order of their value as aids in teaching:

	Rank
a. Charts	
b. Pictures, lantern slides, moving films	
c. Models	
d. Microscopes	
e. Fresh material	
1. Plant	
2. Animal	
f. Preserved material	
1. Plant	
2. Animal	

21. Do you think sex education a desirable part of the biology course?________
    Should it be taught as a separate topic?________
22. Please list four biologic experiments which you think will develop resourcefulness or initiative.
    (1)________________ (2)________________
    (3)________________ (4)________________
23. In your opinion what percentage of the biology laboratory work should be:
    (1) To test for something (as for starch in seeds)?________
    (2) To experience something (as taking own pulse rate; planting seeds and watching their development; caring for animals as a food experiment on rats)?________
    (3) In making drawings and writing up the notebook?________
    (4) Careful observation and write-up of natural objects?________
24. Will you please comment briefly on what you think biological notebook work does contribute to the larger objectives of education given below and what you think it should contribute.

	1. Does Contribute	2. Should Contribute
a.	Health	
b.	Worthy home membership	
c.	Citizenship	
d.	Vocation	
e.	Worthy use of leisure time	
f.	Ethical character	
g.	Command of fundamental processes	

APPENDIX B

(Questionnaire to Principals—Form B)

BIOLOGY NOTEBOOK WORK

Please fill in the following form from your files pertaining to biology work in 1927–28. Use back of sheet if more space is necessary.

1. Pupil's name____________________ Freshman?____ Sophomore?____ Senior?____ (Check classification)
2. List the subjects taken by the above named pupil last year with grades in each subject.

1st Semester Subject	Grade	2nd Semester Subject	Grade

3. List all the biology grades in the above pupil's class or, if you prefer, state the number in the class and give the average grade of all.
   1st Semester____
   2nd Semester____
4. How many biology teachers are there in your school? Men__________ Women________
5. Give the teaching schedule of the above named pupil's teacher.

1st Semester Period	Subject	2nd Semester Period	Subject
1		1	
2		2	
3		3	
4		4	
5		5	
6		6	
7		7	
8		8	

6. Underscore the college degrees the biology teacher holds.
A.B.—B.S.—A.M.—M.S.—Ph.D.—__________ __________ __________
7. State the experience of the above biology teacher in years, including last year.__________ How many years in your school?__________
8. Give the total enrollment of your high school last year:
1st Semester__________ 2nd Semester__________
9. How many pupils last year were enrolled in:—

1st Semester	2nd Semester
Biology__________	Biology__________
Zoology__________	Zoology__________
Botany__________	Botany__________
Physiology__________	Physiology__________

10. How many weeks in your school year?__________
11. What textbook was used in biology last year?__________
12. What biology laboratory manual was used last year?__________
13. Is the biology laboratory and recitation work done in the same room? __________ Is this room equipped as a laboratory?__________ Do you have a separate laboratory and classroom for work in biology? __________
14. Give length of laboratory period in minutes__________ Of recitation period__________ Number of laboratory periods per week__________ Recitation periods per week__________
15. How many compound microscopes in the biological laboratory?__________
16. Is there a projection lantern?__________ A moving picture machine? __________ About how often is the projection lantern used?__________ The moving picture machine?__________
17. Are live animals kept in the biology laboratory?__________ Is there an aquarium in the biology room?__________
18. Please send us a copy of your high school course of study. If you cannot send us a copy, tell us whom we may ask for one and the price to send for it.
Name__________ Address__________ Price__________
19. Did the biology teacher require the notes on laboratory work to be written up during the laboratory period?__________ Could the pupils take notes during the laboratory period and write up the notes in the notebook afterwards?__________ How much of the biology grade was given on the biology laboratory notebook?__________
20. Did the biology class take field trips?__________ About how many field trips were taken last year?__________ Did the teacher go along on field trips?__________
21. How many experiments are required in biology laboratory work?__________
22. How many pupils failed to pass in biology last year? 1st Semester?__________ 2nd Semester?__________ How much science is required for a high school diploma?__________ What science subjects are required?__________ Is biology a required subject?__________
23. How many of the pupils in biology took Regents Examination last year?

January examination?__________ How many passed?__________ June examination?__________ How many passed?__________

24. What was this pupil's mental rating?__________ What intelligence test?__________ When given?__________ What was this pupil's achievement rating?__________ What test?__________ When given? __________

APPENDIX C

(Questionnaire to Students—Form C)

BIOLOGY NOTEBOOK WORK

Please answer, in pencil, the following questions concerning your biology work. If additional space is required use the back of the sheet.

1. Your name____________________ Address____________________
   Freshman____ Sophomore____ Junior____ Senior____ (Check your present classification)
2. You studied biology last year. Which did you enjoy most? The laboratory work?__________ The recitation?__________ The writing of the notebook?__________
3. Which do you think enabled you to understand biology the best? The recitation?__________ The laboratory work?__________ The writing of the notebook?__________
4. How much time was given to the recitation? Periods per week?________ Laboratory work? Periods per week?__________ How much of the recitation time was spent in writing up your notebook?__________ How much of the laboratory time?__________
5. What part of biology did you like best? (Write 1 after the topic you liked best; 2 after the topic you liked second best; and so on.)
   Plants?__________ Animals?__________ Health studies?__________
   Human physiology?__________ Chemical phase?__________ Any others?__________
6. Did your biology class go on any field trips last year?__________ How many field trips?__________ What did you do on the field trips?__________
   ____
   ____
7. What subject in your high school course has been the most interesting?
   ____
   ____
8. List the subjects you took in your high school course last year with the final grade in each. (See directions below for Rank.)

First Semester			Second Semester		
Subject	Final Grade	Rank	Subject	Final Grade	Rank

Which of the above subjects took the most time to prepare?__________ __________ Place 1 on the line after this subject under Rank and mark the others in order of the amount of time required for the preparation, as 2 after the one that comes next in time, and so on.

9. Of the exercises written in your notebook, how many did you do yourself?__________ How many were done by the teacher?__________ How many were done by a group of pupils working together?__________
10. Do you expect to go to college?__________ A coeducational college? __________ A women's college?__________ A college for men?__________
11. List any specific occasions you have had in which you used the information you acquired in your biology work last year.__________
12. Have you taken mental tests?__________ What ones?__________ When?__________ What was your rating?__________ Have you taken achievement tests? __________ What ones?__________ When?__________ What was your rating?__________
13. How many pupils were in your biology class? 1st Semester?__________ 2nd Semester?__________
14. What other topics, if any, do you think your class should have studied in biology?__________ What topics, if any, do you think could just as well have been left out?__________
15. Did you write a term paper? 1st Semester__________ 2nd Semester? __________ What topics? (1)__________ (2)__________ Did any pupils in your class make special reports on any topics?__________ What topics?__________
16. Did you work out any individual projects or problems in biology?__________ What project or problem? 1st Semester__________ What project or problem? 2nd Semester__________ Did any pupils in your class work out special problems or projects in biology?__________ What projects or problems?__________

I shall appreciate very much any other comments you feel like making in regard to high school biology work, especially the notebook phase. (Use the back of this sheet.)

APPENDIX D

LIST OF COLLEGES AND UNIVERSITIES COOPERATING

The colleges and universities to which a letter was sent for specific information regarding the submission of a notebook when offering a unit of biology for admission credit.

Colleges and Universities in New York State

Adelphi College, Brooklyn, N. Y.
Bennington College, New York, N. Y.
Clarkson College of Technology, Potsdam, N. Y.
College of the City of New York, New York, N. Y.
College of New Rochelle, New Rochelle, N. Y.
Columbia University, New York, N. Y.
Concordia Collegiate Institute, Bronxville, N. Y.
Cornell University, Ithaca, N. Y.
Elmira College, Elmira, N. Y.
Fordham University, New York, N. Y.
Hamilton College, Clinton, N. Y.
Hartwick College, Oneonta, N. Y.
Hobart College, Geneva, N. Y.
Hunter College of the City of New York, New York, N. Y.
Houghton College, Houghton, N. Y.
New York University, New York, N. Y.
Niagara University, Niagara Falls, N. Y.
Rensselaer Polytechnic Institute, Troy, N. Y.
Sarah Lawrence College, Bronxville, N. Y.
St. Bonaventure's College, Allegheny, N. Y.
St. Francis College, Brooklyn, N. Y.
St. Lawrence University, Canton, N. Y.
St. Stephen's College, Annandale, N. Y.
Skidmore College, Saratoga Springs, N. Y.
Syracuse University, Syracuse, N. Y.
University of Buffalo, Buffalo, N. Y.
University of Rochester, Rochester, N. Y.
Vassar College, Poughkeepsie, N. Y.
Wells College, Aurora, N. Y.
William Smith College, Geneva, N. Y.

New England Colleges and Universities

Brown University, Providence, R. I.
Dartmouth College, Hanover, N. H.
Harvard University, Cambridge, Mass.
Mount Holyoke College, South Hadley, Mass.
Simmons College, Boston, Mass.
Smith College, Northampton, Mass.
Radcliffe College, Cambridge, Mass.
University of New Hampshire, Durham, N. H.
University of Vermont, Burlington, Vt.
Yale University, New Haven, Conn.

New Jersey Colleges and Universities

Princeton University, Princeton, N. J.
Rutgers University, New Brunswick, N. J.
Seton Hall College, South Orange, N. J.
Upsala College, East Orange, N. J.

Mid-western and Other Colleges and Universities

Baylor College for Women, Bolton, Texas
Colorado State Teachers College, Greeley, Colo.
Ohio State University, Columbus, Ohio
University of Illinois, Urbana, Ill.
University of Chicago, Chicago, Ill.
Stanford University, Stanford University, Calif.
University of Pennsylvania, Philadelphia, Pa.

APPENDIX E

LETTER TO COOPERATING SCHOOLS

THE UNIVERSITY OF THE STATE OF NEW YORK
THE STATE DEPARTMENT OF EDUCATION
EXAMINATIONS AND INSPECTIONS DIVISION

COOPERATIVE STUDY OF NOTEBOOK PRACTICES IN ELEMENTARY BIOLOGY, PHYSICS AND CHEMISTRY

Your school has been selected as one of fifty representative schools from which notebooks are to be submitted to the department this month for an intensive study of the best practices. Attention this year will be limited to notebooks in elementary biology, physics and chemistry, subjects in which it is believed practices are more nearly standardized.

This undertaking is in the nature of a cooperative project between the teacher and the department, working toward improvement of practices in notebook work. Each teacher, therefore, is requested to hold himself or herself in readiness to answer inquiries about any points which may be raised.

In making your selection, choose a single book which you might designate as "good" to "better," but *not* the best.

The books should be labeled with subject, teacher's name and address, school and address, number of weeks in the course, number of experiments performed, name of manual and author, name of text and author, name of principal or head of department and summer address, name of student and home address, and index of experiments.

All books will be carefully preserved. If it is desired to have the notebook returned, make a statement to that effect after the name of the student.

Forward notebooks so as to reach the department not later than June 28, 1928.

CLARENCE E. BAER
State Supervisor of Science

Approved June 13, 1928
AVERY W. SKINNER
Director